Lecture Notes in Computer Science 5880

Commenced Publication in 1973
Founding and Former Series Editors:
Gerhard Goos, Juris Hartmanis, and Jan van Leeuwen

T0223377

Stefano Spaccapietra   Lois Delcambre (Eds.)

Journal on
Data
Semantics XIV

 Springer

Volume Editors

Stefano Spaccapietra
École Polytechnique Fédérale de Lausanne
EPFL-IC, Database Laboratory
1015 Lausanne, Switzerland
E-mail: stefano.spaccapietra@epfl.ch

Lois Delcambre
Portland State University
Department of Computer Science
Portland, OR 97207-0751, USA
E-mail: lmd@cs.pdx.edu

CR Subject Classification (1998): I.2.4, H.3.5, H.2.8, I.2.3, H.2.3, H.3.3, D.3

ISSN   0302-9743 (Lecture Notes in Computer Science)
ISSN   1861-2032 (Journal on Data Semantics)
ISBN-10   3-642-10561-0 Springer Berlin Heidelberg New York
ISBN-13   978-3-642-10561-6 Springer Berlin Heidelberg New York

springer.com

© Springer-Verlag Berlin Heidelberg 2009
Printed in Germany

Typesetting: Camera-ready by author, data conversion by Scientific Publishing Services, Chennai, India
Printed on acid-free paper   SPIN: 12793891   06/3180   5 4 3 2 1 0

The LNCS Journal on Data Semantics

The journal aims to provide a highly visible dissemination channel for remarkable work that in one way or another addresses research and development on issues related to data semantics. The target domain ranges from theories supporting the formal definition of semantic content to innovative domain-specific applications of semantic knowledge. We expect such a publication channel to be of the highest interest to researchers and advanced practitioners working on the Semantic Web, interoperability, mobile information services, data warehousing, knowledge representation and reasoning, conceptual database modeling, ontologies, and artificial intelligence.

Topics of relevance to this journal include:

- Semantic interoperability, semantic mediators
- Ontologies
- Ontology, schema and data integration, reconciliation and alignment
- Multiple representations, alternative representations
- Knowledge representation and reasoning
- Conceptualization and representation
- Multimodel and multiparadigm approaches
- Mappings, transformations, reverse engineering
- Metadata
- Conceptual data modeling
- Integrity description and handling
- Evolution and change
- Web semantics and semistructured data
- Semantic caching
- Data warehousing and semantic data mining
- Spatial, temporal, multimedia and multimodal semantics
- Semantics in data visualization
- Semantic services for mobile users
- Supporting tools
- Applications of semantic-driven approaches

These topics are to be understood as specifically related to semantic issues. Contributions dealing with the semantics of data may be considered even if they are not covered by the topics in the list.

Previous Issues

JoDS I: Special Issue on Extended Papers from 2002 Conferences, Spaccapietra, S., March, S., Aberer, K. (Eds.), LNCS 2800, December 2003

JoDS II: Special Issue on Extended Papers from 2003 Conferences, Spaccapietra, S., Bertino, E., Jajodia, S. (et al.) (Eds.), LNCS 3360, December 2004

JoDS III: Special Issue on Semantic-Based Geographical Information Systems, Spaccapietra, S., Zimanyi, E. (Eds.), LNCS 3534, August 2005

JoDS IV: Normal Issue, Spaccapietra, S. (Ed.), LNCS 3730, December 2005

JoDS V: Special Issue on Extended Papers from 2004 Conferences, Spaccapietra, S., Atzeni, P., Chu, W.W. (et al.) (Eds.), LNCS 3870, February 2006

JoDS VI: Special Issue on Emergent Semantics, Spaccapietra, S., Aberer, K., Cudré-Mauroux, P. (Eds.), LNCS 4090, September 2006

JoDS VII: Normal Issue, Spaccapietra, S. (Ed.), LNCS 4244, November 2006

JoDS VIII: Special Issue on Extended Papers from 2005 Conferences, Spaccapietra, S., Atzeni, P., Fages, F. (et al.) (Eds.), LNCS 4830, February 2007

JoDS IX: Special Issue on Extended Papers from 2005 Conferences (continued), Spaccapietra, S., Atzeni, P., Fages, F. (et al.) (Eds.), LNCS 4601, September 2007

JoDS X: Normal Issue, Spaccapietra, S. (Ed.), LNCS 4900, February 2008

JoDS XI: Special Issue on Extended Papers from 2006 Conferences, Spaccapietra, S. (Ed.), LNCS 5383, December 2008

JoDS XII: Normal Issue, Spaccapietra, S. (Ed.), Vol. 5480, 2009

JoDS XIII: Special Issue "Semantic Data Warehouses", Spaccapietra, S., Zimanyi, E., Song, I.-Y. (Eds.), Vol. 5530, 2009

JoDS Volume XIV

This volume of JoDS consists of five papers, selected on the basis of two rounds of reviews from an initial set of 21 full papers, following the submission of 37 abstracts. The papers were received in response to a call for contributions issued in September 2008. We thank the authors for all their work and for their interest in JoDS. We thank the Editorial Board and the reviewers for their detailed and insightful reviews.

This is the first volume of JoDS for which I have served as Editor. I gratefully acknowledge the cooperation of the Editorial Board and the untiring assistance from Professor Stefano Spaccapietra, Editor-in-Chief of JoDS, as I worked through this process.

Lois Delcambre
Editor, Issue XIV
Co-editor-in-Chief, JoDS with
Stefano Spaccapietra, Editor-in-Chief, JoDS

Reviewers

We are grateful to the external reviewers listed below; they assisted the Editorial Board with the reviewing of papers for this volume.

Jesús Bermúdez, Basque Country University, Spain

Jérôme David, Laboratoire d'Informatique de Grenoble, Pierre Mendes France University, and INRIA, France

Ernesto Jiménez-Ruiz, Universitat Jaume I, Castellon, Spain

Markus Kirchberg, Institute for Infocomm Research, Agency for Science, Technology and Research, Singapore

Gorka Marcos Ortego, Visual Communication and Interaction Technologies Centre, Spain

Andrea Maurino, Università di Milano – Bicocca, Italy

Héctor Pérez-Urbina, Computing Laboratory, University of Oxford, UK

Livia Predoiu, University of Mannheim, Germany

Chan Le Duc, Laboratoire d'Informatique de Grenoble and INRIA, France

Titi Roman, Semantic Technology Institute, University of Innsbruck, Austria

Ioan Toma, Semantic Technology Institute, University of Innsbruck, Austria

Table of Contents

Defining the Semantics of Conceptual Modeling Concepts for 3D Complex Objects in Virtual Reality

Olga De Troyer, Wesley Bille, and Frederic Kleinermann

Vrije Universiteit Brussel
Research Group WISE, Pleinlaan 2
1050 Brussel
Belgium
olga.detroyer@vub.ac.be,
wesley.bille@skynet.be,
frederic.kleinermann@vub.ac.be
http://wise.vub.ac.be/

Abstract. Virtual Reality (VR) allows creating interactive three-dimensional computer worlds in which objects have a sense of spatial and physical presence and can be manipulated by the user as such. Different software tools have been developed to build virtual worlds. However, most tools require considerable background knowledge about VR and the virtual world needs to be expressed in low-level VR primitives. This is one of the reasons why developing a virtual world is complex, time-consuming and expensive. Introducing a conceptual design phase in the development process will reduce the complexity and provides an abstraction layer to hide the VR implementation details. However, virtual worlds contain features not present in classical software. Therefore, new modeling concepts, currently not available in classical conceptual modeling languages, such as ORM or UML, are required. Next to introducing these new modeling concepts, it is also necessary to define their semantics to ensure unambiguousness and to allow code generation. In this paper, we introduce conceptual modeling concepts to specify complex connected 3D objects. Their semantics are defined using F-logic, a full-fledged logic following the object-oriented paradigm. F-logic will allow applying reasoners to check the consistency of the specifications and to investigate properties before the application is actually built.

Keywords: Virtual Reality, F-logic, semantics, complex objects, conceptual modeling, formal specifications, VR-WISE.

1 Introduction

Virtual Reality (VR) is a technology that allows creating interactive three-dimensional (3D) computer worlds (virtual worlds, also called Virtual Environments or VE's) in which objects have a sense of spatial and physical presence

S. Spaccapietra, L. Delcambre (Eds.): Journal on Data Semantics XIV, LNCS 5880, pp. 1–36, 2009.

and can be manipulated by the user as such. VR has gained a lot of popularity during the last decennia due to games and applications such as Second Life [1]. A lot of different software tools have been developed which allow building VE's. However, most tools require considerable background knowledge about VR technology. The development of a VE directly starts at the level of the implementation. A developer needs to specify the VE using the specialized vocabulary of the VR implementation language or framework used. Therefore, when creating a VE the objects from the problem domain have to be translated into VR building blocks (such as textures, shapes, sensors, interpolators, etc.), which requires quite some expertise. This makes the gap between the application domain and the level at which the VE needs to be specified very large, and makes the translation from the concepts in the application domain into implementation concepts a very difficult issue. This is one of the reasons why developing a VE is complex, time-consuming and expensive.

In different domains such as Databases, Information Systems and Software Engineering, conceptual modeling has been used with success to support and improve the development process. The term conceptual modeling denotes the activity of building a model of the application domain in terms of concepts that are familiar to the application domain experts and free from any implementation details. Conceptual modeling has less been used in 3D modeling and VR. However, like for these other domains, introducing a conceptual design phase in the development process of a VR application could be useful to improve and support the development of VE's. It will reduce the complexity and can provide an abstraction layer that hides the specific VR jargon used. In this way, no special VR knowledge will be needed for making the conceptual design of a VE and also non-technical people (like the customer or the end-user) can be involved in the development process. A conceptual model will improve the communication between the developers and the other stakeholders. In addition, by involving the customer more closely in the design process of the VE, earlier detection of design flaws is possible. All this could help in realizing more VR applications in a shorter time.

However, conceptual modeling for VR poses a lot of challenges as VE's involve a number of aspects, not present in classical software or information systems. VE's are 3D worlds composed of 2D and 3D objects and often deal with 3D complex objects for which the way the parts are connected will influence the way the complex objects can behave (i.e. *connected complex objects*). Furthermore, to realize dynamic and realistic worlds, objects may need complex (physical) behaviors. This requires new modeling concepts, currently not available in classical conceptual modeling languages, such as ORM [3] [4] or UML [2]. Next to introducing new modeling concepts, it is also necessary to define their semantics. Defining the semantics of the modeling concepts will allow for unambiguous specifications. Unambiguousness is important from two perspectives. Firstly, if the semantics of the modeling concepts are clear, the models created with these modeling concepts will be unambiguous and there will be no discussion between different stakeholders about their meaning. Secondly, unambiguousness is also needed from the perspective of the code generation; otherwise it will not be possible to automatically generate code.

In this paper, we introduce conceptual modeling concepts to specify connected complex 3D objects, as well as their semantics. These modeling concepts are part of the VR-WISE approach. VR-WISE (Virtual Reality - With Intuitive Specifications Enabled) [5] [6] [7] [8] [9] is a conceptual model-based approach for VR-application development. The semantics of the modeling concepts presented here are defined formally using F-logic, a full-fledged logic following the object-oriented paradigm.

The rest of this paper is structured as follows. Section 2 will provide the background. It includes an introduction to VR (section 2.1), VR-WISE (section 2.2), and F-logic (section 2.3). In section 2.1, we will briefly discuss the different components of a VR application as well as how VR applications are developed these days. In section 2.2, we will discuss the limitations of current conceptual modeling techniques with respect to the modeling of VR, and briefly introduce VR-WISE, the conceptual modeling approach developed for VR. In section 3, we discus the conceptual modeling of complex connected 3D objects and introduce the related conceptual modeling concepts in an informal way. Next, in section 4, the semantics of these modeling concepts will be defined. In section 5 we will discus related work. Section 6 concludes the paper and points out further work.

2 Background

In this section, we introduce some background material. In section 2.1, we briefly discuss the different components of a VR application as well as how VR applications are developed these days. In section 2.2, we discuss the limitations of current conceptual modeling techniques with respect to the modeling of VR, and briefly introduce VR-WISE, the conceptual modeling approach developed for VR. In section 2.3, F-logic is introduced.

2.1 VR

There are many definitions of Virtual Reality (VR) [10] [11]. For the context of this research, VR is defined as a three-dimensional computer representation of a space in which users can move their viewpoints freely in real time. We therefore consider the following cases being VR: 3D multi-user chats (Active Worlds [12], first person 3D videogames (Quake [13]) and Unreal tournament ([14]), and 3D virtual spaces on the Web (such as those created with VRML [15], and X3D [16]). In this section, we will define the main components of a VR application (VE) and then briefly review how a VE is developed today.

Main components of a VR application. A VE is made of different components [10], which can be summarized as:

1) **The scene and the objects.** The scene corresponds to the environment (world) in which the objects are located. It contains lights, viewpoints and cameras. Furthermore, it has also some properties that apply to all the objects located inside the VE, e.g., gravity. The objects have a visual representation with color and material properties. They have a size, a position, and an orientation.
2) **Behaviors** .The objects may have behaviors. For instance, they can move, rotate, change size and so on.
3) **Interaction.** The user must be able to interact with the VE and its objects. For instance, a user can pick up some objects or he can drag an object. This may be achieved by means of a regular mouse and keyboard or through special hardware such as a 3D mouse or data gloves [10].
4) **Communication.** Nowadays, more and more VE are also collaborative environments in which remote users can interact with each other. To achieve this, network communication is important.
5) **Sound.** VR applications also involve sound. Some research has been done over the last ten years in order to simulate sound in a VE.

Developing a VE. The developing of the different components of a VE is not an easy task and during the last fifteen years, a number of software tools have been created to ease the developer's task. These tools can be classified into authoring tools and software programming libraries.

Authoring tools. Authoring tools allow the developer to model the static scene (objects and the scene) without having to program. Nevertheless, they assume that the developer has some knowledge of VR and some programming skills to program behaviors using scripting languages. Different authoring tools may use different scripting languages. The most popular authoring tools are 3D Studio Max [20], Maya [21], MilkShape 3D [22], various modelers such as AC3D [23] and Blender [24]. If the developer is developing for a certain file format, he needs to pay attention to the file formats supported by the authoring tool.

Programming Libraries. With programming libraries a complete VE can be programmed from scratch. Among the existing libraries, there is Performer [25], Java3D [17], X3D toolkit written in C++ [26] or Xj3D [27] written on top of Java3D. To use such a library, good knowledge of programming and a good knowledge of VR and computer graphics are required. It is also possible to use a player that, at run-time, interprets a 3D format and build the VE. VRML [15] and X3D [16] are 3D formats that can be interpreted by special players through a Web browser. Examples of such players are the Octaga player [18] and the Flux player [19].

We will not discus here how the other components of a VE (behavior, interaction, etc.) are development these days; it is not directly relevant for the rest of the paper. In any case, we can conclude that although there are quite a number of tools to help a developer to build the scene of a VE, until now, the general problem with these tools and formats is that they are made for VR specialists or at least for people having programming skills and background in computer graphics or VR. In addition, there is also no well-accepted development method for VR.

Most of the time, a VR-expert meets the customer (often the application domain expert) and tries to understand the customer's requirements and the domain for which the VE is going to be built. After a number of discussions, some sketches are made and some scenarios are specified. Then, the VR-expert(s) start to implement. In other words, the requirements are almost directly translated into an implementation. This way of working usually result into several iterations before the result reaches an acceptable level of satisfaction for the customer. Therefore, the development process is time consuming, complex and expensive.

2.2 VR-WISE

Introducing a conceptual design phase in the development process of a VR application can help the VR community in several ways. As conceptual modeling will introduce a mechanism to abstract from implementation details, it will reduce the complexity of developing a VE and it avoids that people need a lot of specific VR knowledge for such a conceptual design phase. Therefore, also non-technical people (like the customer or the end-user) can be involved and this will improve the communication between the developers and the other stakeholders. In addition, by involving the customer more closely in the design process of the VE, earlier detection of design flaws is possible. And finally, if the conceptual models describing the VR system are powerful enough, it may be possible to generate the system (or at least large parts of it) automatically.

Several general-purpose conceptual modeling languages exist. Well-know languages are UML [2], ER [28] and ORM [3] [4]. ER and ORM were designed to facilitate database design. Their main purpose is to support the data modeling of the application domain and to conceal the more technical aspects associated with databases. UML is broader and provides a set of notations that facilitates the development of a complete software project. To a certain extend, UML, ORM and ER could be used to model the static structure of a VR application (i.e., the scene and the objects), however, all are lacking modeling concepts in terms of expressiveness towards VR modeling. For example, they do not have built-in modeling concepts for specifying the position and orientation of objects in the scene or for modeling connected objects using different types of connections. Although, it is possible to model these issues using the existing modeling primitives, this would be tedious. E.g., each time the modeler would need a particular connection he would have to model it explicitly, resulting in a lot of "redundancy" and waste of time. In addition, the models constructed in this way would not be powerful enough to use them for code generation because the necessary semantics for concepts like connections would be lacking. Furthermore, neither ORM nor ER provides support for modeling behavior.

It could be possible to extend these general-purpose modeling languages with new modeling concepts to enable VR modeling. However, another approach regarding this problem is the creation of a Domain Specific Modeling Language. We have opted for this last approach because we want to have a modeling language, as well as a modeling approach, that is easy and intuitive to use also for non VR-experts. The modeling approach taken by a general-purpose language such as UML is very close to the way software is implemented by means of OO programming languages. The use of (an extended) UML would also force the use of its modeling paradigm. It is our opinion, that for certain aspects of VR, this would not be the best solution. Therefore, we have developed VR-WISE, a domain specific modeling approach for the development of VE's.

VR-WISE includes a conceptual specification phase. During this conceptual phase, conceptual specifications (so-called conceptual models) are created. Such a conceptual specification is a high-level description of the VE, the objects inside the environment, the relations that hold between these objects and how these objects behave and interact with each other and with the user. These conceptual specifications must be free from any implementation details. Therefore, the approach offers a set of high-level modeling concepts (i.e. a modeling language) for building these conceptual specifications. As indicated, we require that these modeling concepts are very intuitive, so that they can be used, or at least be understood, by different stakeholders. This means that the vocabulary used, should be familiar to most of its users. Because we also opted for a model-driven approach, the expressive power of the different modeling concepts must be sufficient to allow code generation from the models.

The conceptual specification consists of two levels since the approach follows to some degree the object-oriented (OO) paradigm. The first level is the *domain specification* and describes the *concepts* of the application domain needed for the VE (comparable to object types or classes in OO design methods), as well as possible relations between these concepts. In the overall example that we will use, we will consider a VE containing virtual mechanical robots. This VE could be used to illustrate the working of those robots. For such an application, the domain specification could contain concepts such as Robot, WeldingRobot, LiftRobot, Controller, WorkPiece, Box, and relations such as "a Robot is driven-by a Controller". Concepts may have *properties* (attributes). Next to properties that may influence the visualization of the concepts (such as height, color, and material) also non-visual properties, like the cost and the constructor of a robot, can be specified. At this conceptual level, we only consider properties that are conceptual relevant. Properties like shape and texture are not necessarily conceptual relevant and may depend on how the object will be visualized in the actual VE. The visualization of the objects is considered in a later phase in the VR-WISE approach. For a VE, behavior is also an important feature. However, the focus of this paper is on modeling concept for complex objects, therefore we will not elaborate on behavior. Details on modeling concepts for behavior can be found in [29] [30] [31] [32] [33] [34].

The second level of the conceptual specification is the *world specification*. The world specification contains the conceptual description of the actual VE to be built. This specification is created by instantiating the concepts given in the domain specification. These *instances* actually represent the objects that will populate the VE. In the robot example, there can be multiple Robot-instances and multiple WorkPiece-instances. Behaviors specified at the domain level can be assigned to objects.

Objects in a VE have a *position* and an *orientation* in the scene (defined in a three-dimensional coordinate system). Although it is possible to specify the position of the instances in a scene by means of exact coordinates and the orientation by means of angles, we also provide a more intuitive way to do this (more suitable for non-technical persons). If you want to explain to somebody how the robot room should look like, you will not do this in term of coordinates. Instead you will say that: "Two welding robots are in front of each other at a distance of one meter. A lift robot is left of each welding robot, and a box is placed on the platform of each lift robot". In such an explanation, spatial relations are used to describe the space. As spatial relations are also used in daily life, they provide a good intuitive way to specify a scene. Therefore, they are available as modeling concepts. Note that although the use of spatial relations may be less exact than coordinates, they are exact enough for a lot of applications. A *spatial relation* specifies the position of an object relative to some other object in terms of a direction and a distance. The following directions may be used: *left, right, front, back, top,* and *bottom*. These directions may be combined. However, not all combinations make sense. For example, the combined direction *left top* makes sense, but *left right* doesn't. Spatial relations can be used in the domain specification as well as in the world specification. In the domain specification, the spatial relations are used between concepts and specify default positions for the instances of a concept. The spatial relations currently supported are the most common ones. It is also possible to consider others, like for instance an "inside" relation. Currently, "inside" can be modeled by considering the object in which another object has to be placed as a new scene.

In a similar way, *orientation relations* can be used to specify the orientation of objects. For example, the *orientation by side relation* is used to specify the orientation of an object relative to another object. It specifies which side of an object is oriented towards which side of another object. E.g., one can specify that the front of the instance WeldingRobot1 is oriented towards the backside of the instance WeldingRobot2.

As common for conceptual languages, the conceptual modeling concepts of VR-WISE also have a graphical notation.

2.3   F-Logic

Frame-Logic (F-logic) is a full-fledged logic. It provides a logical foundation for object-oriented languages for data and knowledge representation. F-logic is a frame-based language; the central modeling primitives are classes with properties (attributes). These attributes can be used to store primitive values or to relate classes to other classes. Subclasses are supported. In this section, we provide a brief introduction to F-logic in order to make the paper self-contained. This introduction is based on [35] and [36] to which we refer the interested reader for more details.

Class Signatures. A class signature specifies names of properties and the methods of the class. To specify an attribute definition ⇒ is used, ⇒⇒ is used to express a multi-valued attribute.

The following statement gives the class signature for the class *professor* :

professor[*publications* ⇒⇒ *article;*
       dep ⇒ *department;*
       highestDegree ⇒ *string;*
       highestDegree •→ *"phd"*]

publications ⇒⇒ *article* states that publications is a multi-valued property. *highestDegree* ⇒ *string* states that highestDegree is a property of type string, and *highestDegree* •→ *"phd"* states that it is an inheritable property, which has the effect that each member-object of the class professor inherits this property and its value. E.g., a member *bill* will have the property *highestDegree* with value *phd* by inheritance. An inheritable property is inherited by a subclass. The inheritable property remains inheritable in this subclass while an inheritable property inherited by a member of the class becomes non inheritable.

Class Membership. In F-Logic we use ":" to represent class membership.

mary : professor
cs : department

Note that in F-logic classes are reified, which means that they belong to the same domain as individual objects. This makes it possible to manipulate classes and member-objects in the same language. This way a class can be a member of another class. This gives a great deal of uniformity.

Method Signatures and Deductive Rules. Next to properties, classes can have methods. Consider the following class:

professor[*publications* ⇒⇒ *article;*
        dep ⇒ *department;*
        highestDegree ⇒ *string;*
        highestDegree •→ *phd;*
        boss ⇒ *professor*]

The property *boss* is actually a method without arguments. In F-Logic, there is no essential difference between methods and properties. The method *boss* takes no arguments as input and gives an object of type professor as output. The following statement is the deductive rule defining the method *boss* for objects of the class *professor*.

P[boss → *B]* ← *P : professor* ∧
          D : departement ∧
          P[dep → *D[head* → *B : professor]]*

The previous statement states that when a member *B* of type *professor* is the head of a departement *D* for which a member *P* of type *professor* is working then *B* is the boss of *P*.

It is also possible to create methods that take one or more arguments as input. Syntactically the arguments are included in parentheses and are separated from the method name by the @-sign. However, when the method takes only one argument the parentheses may be omitted. The following statement gives the signature of a method *papers* for the class *professor*. It takes one argument of type *institution* and returns a set-value of type *article*.

professor[papers@institution ⇒⇒ article]

subclassess. "::" is used to represent the subclass relationship. The following statements denote that *employee* is a subclass of *person* and that *professor* is a subclass of *employee*.

employee :: person
professor :: employee

Predicate. In F-logic, predicate symbols can be used in the same way as in predicate logic, for example:

promotorOf(mary, bill)

Queries. Queries can be considered as a special kind of rules, i.e. rules with an empty head. The following query requests all members of the class *professor* working at the department cs:

?- X : professor ∧ X[dep → cs]

3 Conceptual Modeling Concepts for Connected Complex 3D Objects

In section 2, we have presented an overview of VE's and the VR-WISE approach. As already indicated, complex 3D objects, and more in particular connected complex objects, are important in the context of VE's. In our robot example, a welding robot is a complex connected object, composed of a rail, a base, a lower arm, an upper arm, and a welding head (see figure 1). To enable the specification of such concepts, we need dedicated conceptual modeling concepts.

Let's first start with giving an informal definition of a "complex object" in the context of a VE.

Complex objects are built from other simple and/or complex objects. They are composed by connecting two or more simple and/or complex objects. The connected objects are called components. All components keep their own identity and can be manipulated individually. However, manipulating a component may have an impact on the other components of the complex object. The impact depends on the type of connections used.

Looking to this description, the following issues seem to be important. (1) Complex objects are composed of other objects (components), (2) components

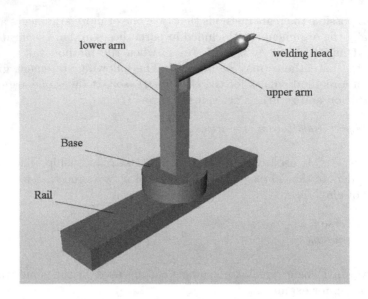

Fig. 1. An illustration of a welding robot

are connected by means of connections and there exist different types of connections, and (3) the motion of a component may be restricted by the connection type used. Let's explain this last issue. Normally an object has six degrees of freedom, three translational and three rotational degrees of freedom. The translational degrees of freedom are translations along the three axes of the coordinate system used while the three rotational degrees of freedom are the rotations around these three axes. The way components of a complex object are connected to each other may restrict the number of degrees of freedom in their displacements with respect to each other. Here, we will discuss three possible connection types, namely over a center of motion, over an axis of motion and over a surface of motion. Other types of connections are possible and can be defined in a similar way. In section 6 on future work, some examples of other types of connections are given. The connection types that we consider here are abstractions from specific connection types usually available in 3D modeling tools (such as ODE [37], PhysX [38], MotionWork [39]). In our conceptual modeling approach, we specify the type of connection between components by means of the so-called *connection relations*. Note that these connection relations can also be used to specify the connection between objects without the purpose of defining a complex object. For example, we can define a connection between a boat and the water surface, but we don't want to consider the boat and the water surface as one complex object.

To support the connection over a center of motion, respectively over an axis of motion and over a surface of motion, we have defined the *connection point relation*, respectively the *connection axis* relation and the *connection surface* relation. We describe them in more detail in the following sections using the welding robot as an example. However, connection types on their own are not sufficient to come to realistic behaviors. To further restrict the behaviors of the connected components, constraints will be used. For instance when you want to state that the components cannot be moved relative to each other. More details on constraints are given in section 3.4. Note that we use the term "components" to refer to the objects involved in a connection relation. However, this does not imply that the use of a connection relation implies defining a complex object. The connection relations are only used to specify connections between objects. Also note that we don't consider here connecting and disconnecting objects at runtime. This is part of the behavior specification, which is outside the scope of this paper.

3.1   Connection Point Relation

A first way of connecting two objects to each other is over a center of motion. In the real world we can find examples of objects connected over a center of motion, e.g., the shoulder of the human body connecting the arm to the torso. In the welding robot example, the welding head is connected to the upper arm over a center of motion to allow the welding head to rotate in different directions. A center of motion means that there is somewhere a point in both components that needs to coincide during the complete lifetime of the connection. We call this point the *connection point*. Connecting two objects over a center of motion removes all three translational degrees of freedom of the components with respect to each other. Specifying a connection point relation implies specifying the connection point for both components. This can be done by means of exact coordinates, however we are looking for a method that is more intuitive for the layman. Therefore, the position of the connection point is specified relative to the *position point* of the object. This is a (default) point in the object that is used to specify the position of the object in the VE, i.e. when an object is positioned at the coordinates (x,y,z) this means that the position point of the object is at position (x,y,z).

Figure 2 shows our graphical notation of the connection point relation. Boxes represent the components; connection relations are represented by a rounded rectangle connecting the two components by means of an arrow. The icon inside the rounded rectangle denotes the type of connection, here a connection point relation. The arrow indicates which component is the source and which is the target. The source should be connected to the target, i.e. in figure 2 component A is connected to component B. Hence, figure 2 can be read as "A is connected by means of a connection point to B". Note that this is not necessarily the same as connecting B to A using the same connection relation. When the composition should be performed if the two composing objects already have a position (e.g., at runtime), it may be necessary to reposition one of the objects. The convention is that the source will be repositioned.

Note that the graphical notation given in figure 2 does not specify the actual connection points. This is done by means of a simple markup language and using the expanded graphical notation (see figure 3). Allowing to hide or to omit

Fig. 2. Graphical notation of connection point relation

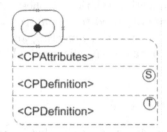

Fig. 3. Extended Graphical notation for a connection point relation

the details of the connections relations is useful as abstraction mechanism in different phases of the design and for different stakeholders. The expanded area has three sub areas. The top area is used to specifying general properties of the connection. Currently, the only possible attribute for the connection point relation is the stiffness. Current values for the stiffness are 'soft', 'medium' or 'hard'. The second and third areas hold the definition of the connection point for the source component respectively for the target component.

The position of a connection point is specified relative to the position point of the component. This is done in terms of zero or more translations of this position point. If no translations are given the connection points coincides with the position point of the object. A translation is specified by a distance and a direction. The distance is expressed by an arithmetic expression and a unit (if no unit is given the default unit will be used). Note that inside the arithmetic expression, object properties can be used. They allow referring to properties of the components, e.g., its width. The direction is given by means of keywords: *left, right, front, back, top* or *bottom*. These directions may be combined. However, not all combinations make sense. A combined direction exists of minimal two and maximal three simple directions. For example, we may use the combined direction *left top*, but *left right* is meaningless. In this way 'translated 2 cm to left' specifies that the connection point is defined as a translation of the position point 2 cm towards the left side. Please note that the position point itself is not changed. The syntax is as follows:

<CPAttributes> ::= [*<stiffness>*]
<stiffness> ::= 'connection stiffness is' *<stiffnessType>*
<stiffnessType> ::= 'soft' | 'medium' | 'hard'
<CPDefinition> ::= 'connection point is position point'*< translation >**
<translation> ::= 'translated' *<distance>* 'to' *<direction>*

<direction> ::= *<A>* *[<B>][<C>]* | *<B>* *[<C>]* | *<C>*
<A> ::= *'front'* | *'back'*
<B> ::= *'left'* | *'right'*
<C> ::= *'top'* | *'bottom'*
<distance> ::= *<arithmetic expression>*

<arithmetic expression> ::=
      <constant> | *<object property>* |
      (< arithmetic expression >) |
      <arithmetic expression> *<operator>* *<arithmetic expression>*

<operator> ::= *+* | *-* | *** |

Figure 4 gives an example of a connection point relation. It is used to connect the welding head to the upper arm of the welding robot. The welding head is the source and the upper arm is the target. Since, the position point of the upper arm is defined in the middle of the upper arm (by default the position point is the centre of the bounding box of the object), the connection point for the upper arm is specified as a translation of the position point over half of the length of the upper arm towards the top. In this way the connection point is exactly on the top of the upper arm. Similar, for the welding head, the connection point (which should be at the bottom of the welding head) is specified as a translation of the position point over half of the length of the welding head towards the bottom.

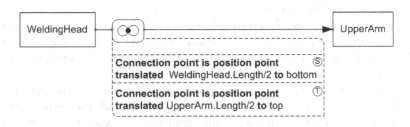

Fig. 4. Example connection point relation

3.2 The Connection Axis Relation

A second way to connect two components is over an axis of motion. A lot of examples of this connection type can be found in the real world. For example: a wheel that turns around an axis, a door connected to a wall, the slider of an old-fashioned typing machine. Actually, an axis of motion means that there is an axis that restricts the displacements of the components with respect to each other in such a way that the connected objects may only move along this axis or around this axis. The axis of motion is called the *connection axis*. A connection by means of a connection axis removes four degrees of freedom leaving only one translational and one rotational degree of freedom.

To specify a connection axis relation between two components, we actually have to specify the connection axis for each of the two components. These two axes need to coincide during the complete lifetime of the connection. Looking for an easy way to specify these axes, we decided to allow a designer to specify an axis as the intersection between two planes. Therefore, three planes through each object are predefined. These are the *horizontal plane, the vertical plane* and *the perpendicular plane*. These planes are illustrated in figure 5.

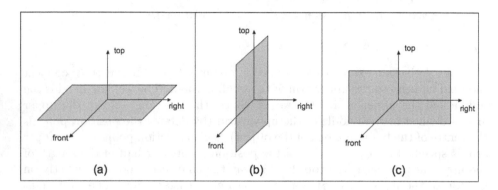

Fig. 5. (a) the horizontal plane; (b) the vertical plane; (c) the perpendicular plane

A connection axis is defined as the intersection between two of these planes. To allow more flexibility, the predefined planes can also be translated or rotated. Each plane may rotate over two possible axes. The horizontal plane may rotate over the left-to-right axis or the front-to-back axis; the vertical plane may rotate over the front-to-back or the top-to-bottom axis; and the perpendicular plane over the top-to-bottom or the left-to-right axis.

Next to define the connection axes it is also necessary to give the initial positions of both components. This is done by specifying for each component a point on its connection axis. These points should coincide. By default this connection point is the orthogonal projection of the position point of the component onto the connection axis. However, our approach also allows the designer to change this default by translating this default point along the connection axis.

The graphical representation of the connection axis relation is similar to that of the connection point relation (see figure 6).

Also in this case, the graphical notation is expandable (see figure 7). The second and third areas are now used for the definition of the connection axis for the source, respectively the target. The syntax is as follows:

$<CAAttributes>$ ::= $[<stiffness>]$
$<stiffness>$ ::= *'connection stiffness is'* $<stiffnessType>$
$<stiffnessType>$ ::= *'soft'* | *'medium'* | *'hard'*
$<CADefinition>$ ::= *'connection axis is intersection of: '*
              $<planeDefinition>$

Fig. 6. Graphical represenation of the connection axis relation

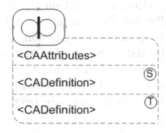

Fig. 7. Expanded graphical representation of the connection axis relation

<*planeDefinition*>
[<*translationPoint*>]

<*planeDefinition*> ::= <*horizontal*> | <*vertical*> | <*perpendicular*>

<*horizontal*> ::=
      '*horizontal plane*' [<*horizontalTrans*>] [<*horizontalRot*>]
<*vertical*> ::=
      '*vertical plane* ' [<*verticalTrans*>] [<*verticalRot*>]
<*perpendicular*> ::=
      '*perpendicular plane* ' [<*perpendTrans*>] [<*perpendRot*>]

<*horizontalTrans*> ::= '*translated* ' <*distance*> '*to* '
('*top*' | '*bottom*')
<*verticalTrans*> ::= '*translated* ' <*distance*> '*to* '
('*left*' | '*right*')
<*perpendTrans*> ::= '*translated* ' <*distance*> '*to* '
('*front*' | '*back*')

<*horizontalRot*> ::=
      '*rotated over* ' ('*frontToBack*' | '*leftToRight*')
'*axis with* ' <*angle*>
<*verticalRot*> ::=
      '*rotated over*' ('*frontToBack*' | '*topToBottom*')
'*axis with* ' <*angle*>
<*perpendRot*> ::=
      '*rotated over*' ('*leftToRight*' | '*topToBottom*')
'*axis with* ' <*angle*>

<angle> ::= <arithmetic expression>

<translationPoint> ::=
     'translation point translated ' <distance> 'to ' <direction>

As an example we show how the base of the welding robot is connected to the rail by means of a connection axis relation to allow the base and the rail to move along this axis. The specification is given in figure 8. For the rail, the connection axis is specified as the intersection of the perpendicular plane with the horizontal plane translated over half of the height of the rail towards the top of the rail. This is illustrated in figure 9. The connection axis on the base is defined as the intersection of the perpendicular plane with the horizontal plane translated to the bottom of the base over half of the height of the base.

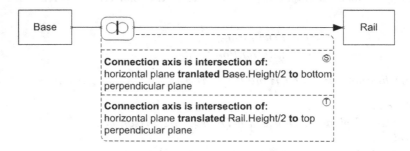

Fig. 8. Example connection axis relation

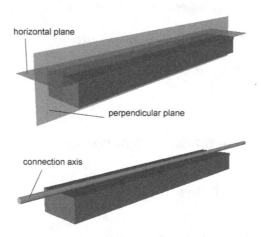

Fig. 9. Illustration of the definition of a connection axis

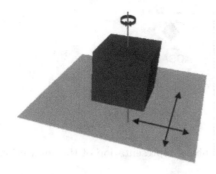

Fig. 10. Degrees of freedom for the connection surface relation

3.3 The Connection Surface Relation

The last way to connect two components to each other that we want to discuss here is over a surface of motion. A real world example of this type of connection is a boat able of floating over a water surface. A surface of motion means that there is a surface that allows the components to move along the directions of this surface. This connection type removes three degrees of freedom. The only degrees of freedom left with respect to each other are the two translational degrees of freedom in the directions of the surface and one rotational degree of freedom around the axis perpendicular to the surface. This is illustrated in figure 10. The surface of motion is called the *connection surface*.

To specify a connection surface relation we actually need to specify the connection surface for each of the components. The connection surfaces of both components need to coincide during the complete lifetime of the connection. To specify these connection surfaces, again we apply the three predefined planes (the horizontal plane, the vertical plane, and the perpendicular plane). For each of the components, the designer selects an initial plane to work with. This plane can be translated and rotated. Similar as for the connection axis relation we also need a connection point to specify the initial position of both components on the connection surface. By default, this point will be the orthogonal projection of the position point of the component on the corresponding connection surface. Also for the connection surface relation, this point can be translated to specify other positions. The graphical representation of the connection surface relation is similar as that of the other connection relations (see figure 11).

Fig. 11. Graphical representation of the connection surface relation

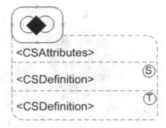

Fig. 12. Extended graphical representation of the connection surface relation

The expanded graphical notation has again three areas: one for specifying the properties of the relation, one area for the specification of the connection surface for the source and one for target. The extended graphical notation is illustrated in figure 12.

The syntax is as follows:

$<$*CSDefinition*$>$ ::= *'connection surface is: '*
  $<$*planeDefinition*$>$ [$<$*CSConnectionPoint*$>$]

$<$*CSConnectionPoint*$>$ ::= *'connection point is positioning point '*
  $<$*distance*$>$ *'to '* $<$*direction*$>$
  [*'and '* $<$*distance*$>$ *'to '* $<$*direction*$>$]

In figure 13 a connection surface relation is used to specify that a box placed on the platform of a lift robot should only be able to move over this platform. The connection surface for the box is defined as the default horizontal plane of the box translated towards the bottom of the box; the connection surface for the platform is also the horizontal plane (but this time it is the horizontal plane of the platform itself because the platform is a plane itself and therefore it will coincide with its horizontal plane).

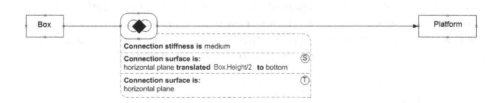

Fig. 13. Connection Surface for lift robot with a platform

3.4 Constraints on Connections

So far we are able to specify connection relations between components. As discussed, these relations impose a limitation on the degrees of freedom of the components with respect to each other. However, this is not always sufficient to come to realistic behaviors. For example, by means of the connection axis relation used to connect a base to its rail, it is still possible to rotate the base and the rail around the connection axis. This is not what we want. We would like to be able to specify that the base should only be able to move along its connection axis. Instead of defining this as yet another special kind of connection relation, we have opted to specify these kinds of restrictions by means of constraints that can be specified on top of the connection relations. For our base-example, a constraint can be attached to the connection axis relation stating that the base may only move over a given distance along its connection axis. A number of constraints are predefined, e.g., the *hinge constraint*, the *slider constraint* and *the joystick constraint*. The names of the constraints are metaphor-based which should make it easier for non-technical persons to understand and remember their meaning. For example, the restriction of the base motion can be expressed by a slider constraint.

A slider constraint can be defined on top of a connection axis relation to restrict the motion to a move along the connection axis. Furthermore, the move can be limited by indicating how much the components may move along the connection axis. Figure 14 illustrates the specification of a slider constraint for the welding robot. The constraint is defined on top of the connection axis relation that connects the base to the rail. The base can move 2,5 units to the left and to the right.

A hinge constraint is also specified on top of a connection axis constraint and restricts the motion to a rotation around the connection axis. It is also possible to indicate limits for this movement. The joystick constraint restricts

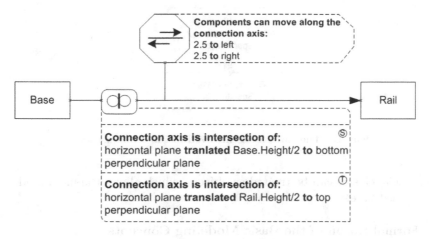

Fig. 14. An example of a slider constraint

the motion of two components connected by means of a connection point relation to a rotation around two perpendicular axes through the connection point. A joystick constraint can also have limits indicating how much the components may rotate around the axes in the clockwise and in the counterclockwise direction. More information on these constraints can be found in [5].

4   Formal Specification of the Modeling Concepts

In this section we will illustrate how the semantics of the modeling concepts introduced in the previous section can be defined rigorously. Due to space limitations, it is not possible to give the complete formalization of the modeling concepts introduced in this paper. For the complete formalization we refer to [5]. We will focus on the principles used and give a representative number of formalizations.

4.1   Principles

To define the semantics of a modeling concept, we will express what its use means for the actual VE. E.g., if we state that two objects are connected by means of a connection axis relation then the semantics of this relation should define the implication for the position and orientation of the components in the VE. To be able to do this, we need a formalism that is able to deal with the instance level. However, because the instance level is defined through the concept level, we also have to deal with the concept level. This means that defining the semantics of the modeling concepts (i.e. the meta-level) requires access to the concept level and to the instance level (see figure 15).

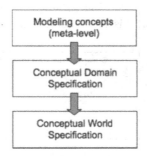

Fig. 15. The three levels involved in the formalization

In F-logic, classes can be treated as objects, which allows the meta modeling that we need here.

4.2   Formalization of the Basic Modeling Concepts

In this section, we formally define some of the basic modeling concepts used in the VR-WISE approach.

Point. A point in a three dimensional space can be given by means of an x, y and z coordinate. Using the F-logic formalism, a point is defined as a class with x, y and z properties of type float.

$point[x \Rightarrow float;$
$\quad y \Rightarrow float;$
$\quad z \Rightarrow float]$

Orientation. Each object in a VE has an orientation. In the VR-WISE approach, each object specified in the world specifications will have a default orientation. This default orientation is illustrated in figure 16. The default orientation of an object can be changed by rotating the object around on or more of the axes of the reference frame used. In order to be able of expressing the orientation of objects, we define the class orientation with properties *frontAngle*, *rightAngle* and *topAngle*, each representing a rotation angle around respectively the front, left and top axis of the global reference frame. In the default situation all rotation angles are 0.

$orientation[frontAngle \bullet\!\!\rightarrow 0;$
$\quad rightAngle \bullet\!\!\rightarrow 0;$
$\quad topAngle \bullet\!\!\rightarrow 0]$

Fig. 16. Default orientation on an object

Line. We have formally defined a line with the following parametric equations:

$$\begin{cases} x = x_0 + ta \\ y = y_0 + tb \\ z = z_0 + tc \end{cases}$$

as follows in F-logic:

$line \, [\, x_0 \Rightarrow float;$
$\quad y_0 \Rightarrow float;$
$\quad z_0 \Rightarrow float;$
$\quad a \Rightarrow float;$
$\quad b \Rightarrow float;$
$\quad c \Rightarrow float \,]$

4.3   Formalization of Concept and Complex Concept

Concept. A concept is the main modeling concept in the domain specification. We have defined a concept in F-logic as a class. So domain concepts are represented as classes in F-logic. Each class representing a domain concept needs to be defined as a subclass of the predefined class *concept*. The class *concept* is defined as follows:

concept[position ⇒ point;
        internalOrientation ⇒ orientation;
        externalOrientation ⇒ orientation]

The properties *position*, *internalOrientation*, and *externalOrientation* are methods. The instances of the concepts are the actual objects in the VE. Objects in a VE have a position and an orientation. In VR-WISE we make use of an internal orientation and an external orientation (explained further on). The corresponding methods in the class concept are inheritable properties and therefore each instance of any subclass will inherit them. These methods can therefore be used to return the position, respectively internal and external orientation of an object. The definitions of these methods are rather elaborated (and therefore omitted), as they need to take into account whether the object is a component of some other object and how its position and orientation has been specified. I.e., as explained earlier, objects can be positioned using exact coordinates but also relative to other objects using spatial relations. Therefore, the position can either be exact or relative. When the designer specifies the position by means of coordinates (a point), the exact position is known and there is no need to calculate the position. However, when the position is given by means of some spatial relations or connection relations, only relative positions are given and the actual coordinates need to be calculated. Also the orientation can be exact (given by means of angles) or relative (given by means of orientation relations).

Because of the use of orientation relations in VR-WISE, each object has been given an internal and an external orientation. The *internal orientation* of an object is used to specify which side of the object is defined as the front, back, left, right, top and bottom side. The *external orientation* of an object is used to specify the orientation of the object in the VE. This works as follows.

By default each object has an own reference frame. This is the coordinate system local to the object where the origin of the coordinate system is the origin of the object (which is also the position point). Figure 17(a) illustrates the default internal orientation for an object. By default, the top/bottom, front/back, and right/left side of an object are defined as in Figure 17(a). It is possible to change this default by means of a rotation of the local reference frame of the object around some of the axes of the global reference frame. (The global reference frame refers to the coordinate system of the VE itself. The origin of the global reference frame is the origin of the VE.) Figure 17(b) illustrates a non-default definition of top/bottom, front/back, and right/left. This is done by specifying an internal orientation which is a 45 degrees counterclockwise rotation of the default internal orientation around the front direction. Note that the object

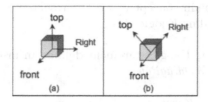

Fig. 17. (a)default internal orientation; (b) internal orientation 45 degrees counterclockwise around front axis

itself is not rotated. Actually, changing the internal orientation only redefined the left-right and top-bottom sides of the object.

An object also has a default external orientation. Figure 18(a) illustrates the default external orientation for an object. The external orientation of an object can be used to change the orientation of the object in the VE. This is done by a rotation of the object around some of the axes of the object's local reference frame, which will result in a rotation of the object itself. This is illustrated in figure 18(b) where the default external orientation is changed by means of a rotation of 45 degrees counterclockwise around the front axis. As you can see, the complete object has been rotated and as such its orientation in the VE is changed.

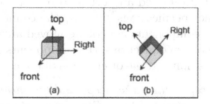

Fig. 18. (a) default external orientation; (b) external orientation 45 degrees counterclockwise around front axis

Example. The following F-logic statement defines the concept WeldingRobot:

WeldingRobot :: concept

The concept *WeldingRobot* may have one or more properties. Suppose it has a weight property with default value 1500 and a color property with default value 'red'. All properties of a concept should be defined as inheritable properties. This way, all subclasses and instances of a concept will inherit the default values. They can be overwritten if necessary for a specific subclass or instance. Let's go back to our example. The properties for WeldingRobot should be defined as follows:

WeldingRobot[weight •→ 1500;
color •→ red]

An instance of a domain concept will be represented in the F-logic as an instance of the corresponding F-logic class.

Example. The following F-logic statement defines an instance *WeldingRobot1* of the domain concept *WeldingRobot*:

WeldingRobot1 : WeldingRobot

To overwrite the default value for color, we use the following statement: *WeldingRobot1[color → green]*

Complex Concept. Remember that a complex concept consists of a number of components. Components can be simple concepts or complex concepts. To specify that a concept is a component of some other concept, a *partOf* property is used.

Let a and b be two concepts, thus a::concept and b::concept. The fact that a is part of b is expressed by adding a property partOf to the concept definition of a: a[partOf ⇒ b]

Each complex concept also has a *reference component*. The reference component is used for the positioning and orientation of the complex concept. The position and orientation of the reference component will also be the position and orientation of the complex concept. In addition, all other components of the complex concept will be positioned and oriented relative to this reference component (according to the specifications given by the connection relations). This does not imply that the components cannot be moved anymore. If it is not forbidden by constraints and/or connections, the components can still be moved. In this case, the relative position and the orientation of the components will change.

Let a and b be two concepts, thus a::concept and b::concept. The fact that a is the reference part of b is expressed by adding a property referencePartOf to the concept definition of a: a[referencePartOf ⇒ b]

Note that when a concept is the reference component of a complex concept, then this concept should also be a component of this complex concept.
Let a and b be two concepts, thus a::concept and b::concept, where a is the reference part of b. Then the following deductive rule holds: a[partOf ⇒ b] ← a[referencePartOf ⇒ b]

The *partOf* property and the *referencePartOf* property are used to define the modeling concept *complexConcept*. It is defined as a subclass of the *concept* class (since a complex concept is also a concept) with properties (methods) *allParts* and *referencePart*

complexConcept :: concept
complexConcept[allParts ⇒⇒ concept;
            referencePart ⇒ concept]

allParts and *referencePart* are methods defined using the following deductive rules:

$$C[allParts \twoheadrightarrow A] \leftarrow C : complexConcept \wedge$$
$$A : concept \wedge$$
$$A[partOf \rightarrow C]$$

$$C[referencePart \rightarrow A] \leftarrow C : complexConcept \wedge$$
$$A : concept \wedge$$
$$A[referencePartOf \rightarrow C]$$

The position (method) for a complex object is defined as follows:

$$A[position \rightarrow P] \leftarrow A:complexConcept \wedge C:concept \wedge P:point \wedge$$
$$C[referencePartOf \rightarrow A] \wedge$$
$$C[position \rightarrow P]$$

Example. The following F-logic statements define a WeldingRobot as a complex concept composed of a Rail, a Base, a LowerArm, an UpperArm, and a WeldingHead. The Rail is used as reference component.

WeldingRobot :: complexConcept
Rail :: concept
Base :: concept
LowerArm :: concept
UpperArm :: concept
WeldingHead :: concept
Rail[partOf ⇒ WeldingRobot]
Base[partOf ⇒ WeldingRobot]
LowerArm[partOf ⇒ WeldingRobot]
UpperArm[partOf ⇒ WeldingRobot]
WeldingHead[partOf ⇒ WeldingRobot]
Rail[referencePartOf ⇒ WeldingRobot]

To create an instance of the complex concept, the following F-logic statements could be used:

WeldingRobot1 : WeldingRobot
Rail1 : Rail
Base1 : Base
LowerArm1 : LowerArm
UpperArm1 : UpperArm
WeldingHead1 : WeldingHead
Rail1[partOf → WeldingRobot1]
Base1[partOf → WeldingRobot1]
LowerArm1[partOf → WeldingRobot1]

$UpperArm1[partOf \rightarrow WeldingRobot1]$
$WeldingHead1[partOf \rightarrow WeldingRobot1]$
$Rail1[referencePartOf \rightarrow WeldingRobot1]$

If we would now like to know the position of the instance $WeldingRobot1$ in the VE, we could use the position method from the *concept* class (using tools like OntoBroker [35] or Flora-2 [36]).

In a similar way, we can define the external and internal orientation of a complex concept. Details are omitted.

So far we have shown how complex concepts are formalized. Note that all methods are working on instances of complex concepts. However, it may also be useful to be able to query the concept level and ask the system about the concepts that are part of some complex concept. Therefore we have overloaded the methods *allParts* and *referencePart* so that they also work on classes:

$$C[allParts \twoheadrightarrow A] \leftarrow C :: complexConcept \land$$
$$A :: concept \land$$
$$A[partOf \Rightarrow C]$$

$$C[referencePart \rightarrow A] \leftarrow C :: complexConcept \land$$
$$A :: concept \land$$
$$A[referencePartOf \Rightarrow C]$$

Concepts themselves don't appear in the VE (only the instances), therefore they don't have a position and neither an orientation. Therefore, there is no need to overload the methods *position*, *externalOrientation* and *InternalOrientation*.

4.4 Formalization of the Connection Relations

In this section we will explain how the connection relations are formalized. Note that the semantics of the connection relations can be considered from two different viewpoints. We can consider the *axiomatic semantics*, which only defines the specific properties (expressed as assertions) of the effect of using the connection relation. However, it is also possible to consider the *operational semantics*, which defines the meaning of the connection relation by the computation it induces when it is used. Expressing the axiomatic semantics is easier as the aspects of the execution are ignored. For example, for a connection axis relation it would be sufficient to state that the two connection axes need to coincide. For the operational semantics, it is also necessary to specify how this should be realized. This makes it much more complex, however, such a semantics has the advantage that the formal specifications can also be used to calculate actual positions and orientations and to reason about this when objects are actual moved or manipulate inside the VE. As we have plans to use our formalization for this purpose, we decided to use the operational semantics.

Using the operational semantics, we also had to distinguish between what we call the *initial semantics* and the *simulation semantics*. On the one hand, a connection relation expresses how two components should be actually connected and in this way defines what this means in terms of the position and orientation of the two components. This is what we will call the initial semantics of the connection relation, as it defines the semantics of the connection relation at the time it is used to connect two components. On the other hand, a connection relation also expresses constraints on the possible motions of the components during the rest of the lifetime of the connection. This is what we will call the simulation semantics.

Because of the complexity introduced by the operational semantics, we will only give the formalization for the connection axis relation and also not completely elaborate all details. Details can be found in [5].

Formalization of the connection relations. Remember that the connection axes are defined using 3 predefined planes: the horizontal plane, the vertical plane and the perpendicular plane. These planes are defined as subclasses of plane, a class without properties. The definition of the horizontal plane is given below. As we have seen earlier, the default horizontal plane can be rotated over the left-to-right axis and over the front-to-back axis. This is expressed by the properties *leftToRightAngle* and *frontToBackAngle* (with default value 0). The position is given by the properties x0, y0 and z0 (default value 0). Translating the horizontal plane along the top-to-bottom axis is expressed by overwriting the z0 property. The vertical plane and the perpendicular plane are defined in a similar way.

horizontal :: plane
horizontal [*leftToRightAngle* $\bullet\!\!\!\to$ *0;*
          frontToBackAngle $\bullet\!\!\!\to$ *0;*
          x_0 $\bullet\!\!\!\to$ *0;*
          y_0 $\bullet\!\!\!\to$ *0;*
          z_0 $\bullet\!\!\!\to$ *0*]

Now we are ready to formalize the connection axis relation. First we will formalize the initial semantics of the connection axis relation. The *connectionAxis Relation* class contains a number of properties to identify for each component the two planes used to define the connection axis (*sourcePlane1, sourcePlane2, targetPlane1,* and *targetPlane2*). There are also a number of properties that specify the connection point for the source and the target component (*sourceTP Dist, sourceTPDir, targetTPDist, tragetTPDir*). Remember that a connection point is specified by a distance and a direction. The *connectionAxisRelation* class also has two properties *sourceAxis* and *targetAxis* which are methods that return the actual axis (a line) which is the connection axis for the source, respectively the target component.

connectionAxisRelation [*sourcePlane1* $\Rightarrow$ *plane;*
                    sourcePlane2 $\Rightarrow$ *plane;*
                    targetPlane1 $\Rightarrow$ *plane;*
                    targetPlane2 $\Rightarrow$ *plane;*

$$sourceAxis \Rightarrow line;$$
$$targetAxis \Rightarrow line;$$
$$sourceTPDist \Rightarrow float;$$
$$sourceTPDir \Rightarrow string;$$
$$targetTPDist \Rightarrow float;$$
$$targetTPDir \Rightarrow string]$$

The fact that two components are connected by means of a connection axis relation is given by means of the *connectTo* property:

Let a and b be two concepts (a:: concept and b:: concept). The fact that a is connected to b by means of a connection axis relation r (r : connectionAxis Relation) is expressed as follows:

$a [connectedTo@b \bullet\!\!\rightarrow r]$

The method *sourceAxis* is defined as follows:

$$C[sourceAxis \rightarrow L] \leftarrow C : connectionAxisRelation \wedge L : line \wedge$$
$$A[connectedTo@B \bullet\!\!\rightarrow C] \wedge$$
$$A[position \rightarrow p] \wedge$$
$$intersectionLine(C.sourcePlane1,$$
$$C.sourcePlane2, M) \wedge$$
$$L : line[x_0 \rightarrow M.x_0 + p.x;$$
$$y_0 \rightarrow M.y_0 + p.y;$$
$$z_0 \rightarrow M.z_0 + p.z;$$
$$a \rightarrow M.a; b \rightarrow M.b; c \rightarrow M.c]$$

The predicate *intersectionLine* is used to state that a line M is the intersection of two planes. The definition of this predicate is omitted. p is the position of the source component. L (the connection axis) express a line through the point p and parallel with M (given by parametric equations).

The method *targetAxis* is defined in a similar way as *sourceAxis* (source should be replaced by target).

Now, we can define the initial semantics of a connection axis relation in terms of the position and orientation of the source component. This is done by means of the predicates *connectedPosition*(A, C, P) and *connectionAxisPos*(A, C, P).

The predicate *connectedPosition*(A, C, P) states that P is the position of a concept A connected via a connection relation to a concept C. The predicate works for A being the source of a connection relation as well as for A being the target. Note that the predicate can be defined in such a way that it applies for whatever type of connection relation used to make the connection between A and C. However, to keep it simple, the definition given here only takes a connection axis relation into account.

$$connectedPosition(A, C, P) \leftarrow A : concept \wedge C : concept \wedge$$
$$P : point \wedge (A [connectedTo@C \bullet\!\!\rightarrow R] \wedge$$
$$R : connectionAxisRelation \wedge$$

$connectionAxisPos(A,\ C,\ P))\ \lor$
$(C\ [connectedTo@A \leftrightarrow R] \land R : connectionAxisRelation \land$
$TPPos(A,\ R,\ P))$

The predicate $connectionAxisPos(A, C, P)$ states that P is the position of A when it is connected via a connection axis relation to C, while the predicate $TPPos$ states that P is the position of A when it is playing the role of target object in the connection axis relation R. Details of this predicate are omitted.

The predicate $connectedOrientation(A, C, E)$ is defined to state that E is the external orientation for A connected via a connection relation to a concept C.

$connectedOrientation(A,\ C,\ E) \leftarrow$
     $A : concept \land C : concept \land E : orientation \land$
     $A\ [\ connectedTo@C \rightarrow R]\ \land$
     $R : connectionAxisRelation \land connectionAxisOrient(A,\ C,\ E)$

The predicate $connectionAxisOrient(A, C, E)$ states that E is the external orientation of A when it is connected via a connection axis relation to C. Details of this predicate are omitted.

So far we have formalized the initial semantics of the connection axis relation. However, we also need to define the meaning of the connection axis relation for the rest of the simulation, the so-called in simulation semantics. In the initial semantics, the source needs to be positioned and oriented according to the target's position and orientation, taking into account the connection axis relation. During the simulation, the difference between source and target is not relevant anymore. When one of the objects moves or changes orientation, the other one has to move with it in such a way that their connection axis relations still coincide.

If two concepts a and b (thus a :: concept and b :: concept) are connected over a connection axis relation R, then for their simulation semantics the position and orientation of a and b must be so that:

$caPosConstraint(a,\ b,\ P_a)\ \land a[position \rightarrow P_a]$
$caPosConstraint(b,\ a,\ P_b)\ \land b[position \rightarrow P_b]$

$connectionAxisOrient(a,\ b,\ E_a)\ \land a[externalOrientation \rightarrow E_a]$
$connectionAxisOrient(b,\ a,\ E_b)\ \land a[externalOrientation \rightarrow E_b]$

The predicate $caPosConstraint(A, C, P)$ states that P is the position of A when it is connected via a connection axis relation to C. Note that this predicate is very similar to the predicate $connectionAxisPos$. However, for the predicate $connectionAxisPos$ the source is positioned along the connection axis according to the specified translation point. Now for the simulation semantics, the translation points are not relevant anymore. Therefore a different predicate was needed. We omit the details of $caPosConstraint$.

5   Related Work

This section will discuss related work. The first subsection reviews a number of academic modeling approaches. The second subsection reviews a number of languages that supports the modeling of rigid-body and the third subsection reviews a number of commercial modeling approaches.

5.1   Academic Modeling Approaches

Kim et al. propose a structural approach for developing VR applications [40]. The approach is called ASADAL/PROTO and it uses Visual Object Specification (VOS). The primary purpose of VOS is to describe physical properties and configuration of physical entities. Spatial constraints can be used to define a structure that is similar to a scene graph. However, there is no support in VOS to describe physical connections and constraints between different objects.

The CODY Virtual Constructor [41] [42] [43] is a system which enables an interactive simulation of assembly processes with construction kits in a virtual environment. The assembly process happens either by means of direct manipulation or by means of natural language. Connections happen by means of predefined points on the graphical objects. These predefined points are called connection ports. When a moved object is in a position so that one of its connection ports is close enough to the connection port of another object, a snapping mechanism will fit the objects together. The core of the CODY architecture is based on a knowledge processing component that maintains two levels of knowledge, namely a geometric level and a conceptual level. For the representation of the knowledge inside the knowledge bases a framebased representation language COAR (Concepts for Objects, Assemblies and Roles) has been developed. The use of natural language offers a very intuitive way of describing an assembly. However, natural language is often ambiguous and incomplete. This means that the outcome of some natural language assembly modeling might not be what the designer wants. Another disadvantage is that the connection ports must be defined in advance. A third disadvantage might occur with very large assemblies where it can be difficult for the designer to find his way through all the connection ports.

The aim of the Open Assembly Model (OAM) [44] is to provide a standard representation and exchange protocol for assembly information. In fact it is defined as an extension to the NIST Core Product Model (CPM) which was presented in [45]. The class Artifact (which comes from the CPM) refers to a product or one of its components. It has two subclasses, namely Assembly and Part. An Assembly is a composition of its subassemblies or parts. OAM has been designed to represent information used or generated inside CAD3-like tools. This enhances the product development across different companies or even within one company. However, OAM is not targeting the modeling of assemblies on a conceptual level. It is targeting an underlying representation of assemblies inside the domain of engineering.

The Virtual Assembly Design Environment (VADE) [46] [47] is a VR-based engineering application that allows engineers to evaluate, analyze, and plan the assembly of mechanical systems. The system utilizes an immersive virtual environment coupled with commercial CAD systems. VADE translates data from the CAD system to the virtual environment. Once the designer has designed the system inside a CAD system, VADE automatically exports the data into the virtual environment. Then, the VR user can perform the assembly. During the assembly process the virtual environment keeps a link with the CAD system. At the end of a VADE session, the design information from the virtual environment is made available in the CAD system. The VADE system is intended for engineers and it is far from high-level.

The Multi-user Intuitive Virtual Environment (MIVE) [48] [49] [50] provides a simple way for objects to constrain to each other without having to use a complete constraint solver. MIVE uses the concept of virtual constraints. Each object in the scene is given predefined constraint areas. These areas can then be used to define so-called offer areas and binding areas. One major disadvantage of the MIVE approach is that for example for the virtual constraints each object in the scene needs predefined areas. Therefore it is difficult to reuse existing virtual objects without adapting them to the MIVE approach.

5.2 Languages Supporting Rigid-Body

Recently, a revision of the X3D architecture [51] and base components specification includes a rigid body physics component (clause 37 of part 1 of the X3D specification). This part describes how to model rigid bodies and their interactions by means of applying basic physics principles to effect motion. It offers various forms of joints that can be used to connect bodies and allow one body's motion to effect another. Examples of joints offered are BallJoint, SingleAxisHingeJoint, SliderJoint or UniversalJoint. Although X3D is sometimes entitled as being high-level, it is still focussed only on what to render in a scene instead of how to render the scene. X3D is still not intuitive for a non-VR expert as for example, he still needs to specify a hinge constraint by means of points and vectors. Furthermore, reasing over X3D specifications is far from easy.

Collada [52] also has a rigid body specification in the same way as X3D. But Collada has been created as an independent format to describe the 3D content that can be read by any software. Therefore, Collada is a format for machine and not really for human and certainly not for describing the modeling of complex objects from a high-level point of view.

5.3 Commercial Modeling Approaches

SimMechanics [53] is a set of block libraries and special simulation features to be used in the Simulink environment. Simulink is a platform for simulation in different domains and model-based design for dynamic systems. It provides an interactive graphical environment that can be used for building models. It contains the elements for modeling mechanical systems consisting of a number of rigid bodies connected by means of joints representing the translational and rotational degrees of freedom of the bodies relative to one another. Although SimMechanics is already on a higher-level of abstraction (than for example physics engine programming), it is still too low-level to be generally usable for application domain experts. Another disadvantage of the approach is that there is no possibility to do some reasoning.

SolidWorks [54] is a 3D computer-aided design (CAD) program in which 3D parts can be created. These 3D parts are made by using several features. Features can be for example shapes and operations like chamfers or fillets. Most of the features are created from a 2D sketch. This is a cut-through of the object which can for example be extruded for creating the shape feature. MotionWorks makes it possible to define mechanical joints between the parts of an assembly inside SolidWorks. MotionWorks contains different families of joints. It is fully integrated into Solid-Works. These tools are not suited for non-experts. The vocabulary used in SolidWorks is really meant for the expert.

3D Studio Max (3ds max) [20] is a 3D modeling software in the category of authoring tools. There are also other tools ([24], Maya[21]) similar to 3ds max and therefore this paper will review 3ds max as a representative of this category. 3ds max provides grouping features which enable the user to organize all the objects with which he is dealing. 3ds max has also another way of organizing objects, namely by building a linked hierarchy. 3ds max provides a number of constraints that can be used to force objects to stay attached to another object. Although 3ds max is intended to create virtual environments without the need for detailed programming, one needs to be an expert in the domain of VR to be able to use an authoring tool like 3ds max. The vocabulary used in the menus and dialogs of such an authoring system is very domain specific. Terms like NURBS, splines or morph are more or less meaningless for a layman.

6   Conclusions and Further Work

In this paper, we have described why conceptual modeling can be important for the field of VR. We also explained the shortcomings of current conceptual modeling languages with respect to VR. Next, we have presented a conceptual modeling approach for VR and we have introduced conceptual modeling concepts to specify complex connected 3D objects. The semantics of the modeling concepts presented are defined formally using F-logic, a full-fledged logic following the object-oriented paradigm. Operational semantics have been defined for the modeling concepts. The use of operational semantics has the advantage of being able to actually calculate positions and orientations of objects and to reason about the specifications. The use of F-logic also allows using a single language for specifying the three different levels involved in our approach, i.e. the meta-level, the concept-level and the instance-level, as well as for querying.

We do not claim that the conceptual modeling concepts presented here are a complete set of modeling primitives for 3D complex objects. One limitation concerning the modeling of connections is that it is not yet possible to define a connection between two components that is a combination of connections, or to combine constraints on connections. This is needed to allow for more powerful connections such as e.g., a car wheel. For this we need actually a combination of two hinge constraints. The problem however is that the motion allowed by one hinge constraint may be in contradiction with the motion allowed by the second hinge constraint, and therefore a simple combination is not sufficient. Next, we also need modeling concepts for specifying so-called contact joints. This type of joints does not really describe a connection but rather a contact between two objects like the gearwheels of a watch that need to roll against each other.

Another limitation is that the approach presented here is only usable for modeling VE's up to a certain level of complexity. E.g., the approach is difficult to use for modeling detailed mechanical assemblies. This is due to the fact that these types of virtual objects require a very high level of detail and also because of the fact that domain specific concepts are needed. However, a layer can be built on top of our approach that pre-defines these necessary domain specific modeling concepts. Such an extension can be made for different domains. However, our approach can always be used for fast prototyping. Prototype tools [55] are developed that allow generating code from the (graphical) conceptual specifications. Afterwards, VR experts using other tools such as VR toolkits may refine the prototype generated from the conceptual specifications.

The formalization given, unambiguously defines the modeling concepts. However, in order to use it for reasoning and consistency checking, an implementation needs to be built. I.e. the conceptual specifications, which are given by a designer using the graphical notation, need to be translated into their corresponding F-logic representation and added to some knowledge bases. Having all the information in F-Logic knowledge bases, we can then use existing F-logic systems (possibly extended with extra features), such as Flora-2, to query the conceptual specifications and to do some reasoning and consistency checking. Currently, we are working on such a reasoning system. It allows specifying a number of domain-independent as well as domain-dependent rules. Examples of domain-independent rules are: a rule to ensure that the partOf relation is anti-symmetric, a rule to detect complex objects that don't has a reference object, or a rule that detects objects that are placed at the same location. Examples of domain-specific rules are: a rule that specifies that all robot-instances should be positioned on the ground, or a rule that each robot should have a controller. More on this can be found in [56].

For future work we plan to extend the implementation to be able to dynamically update the conceptual specifications at run-time. Therefore we need to implement a mechanism so that changes in the actual VE are directly reflected in the logical representation. Such an extension would also allow querying VE's in real time.

Acknowledgment

The work presented in this paper has been funded by FWO (Research Foundation - Flanders) and partially also by IWT (Innovation by Science and Technology Flanders).

References

1. Second Life, http://www.secondlife.com/
2. Fowler, M., Kendall, S.: UML Distilled: a brief introduction to the standard object modeling language. Addison-Wesley Professional, London (1999)
3. Halpin, T.: Conceptual Schema and Relational Database Design. WytLytPub (1999)
4. Halpin, T.: Information Modeling and Relational Databases: From Conceptual Analysis to Logical Design. Morgan Kaufmann, San Francisco (2001)
5. Bille, W.: Conceptual Modeling of Complex Objects for Virtual Environments, A Formal Approach. Ph.D. thesis, Vrije Universiteit Brussel, Brussels, Belgium (2007)
6. Bille, W., De Troyer, O., Kleinermann, F., Pellens, B., Romero, R.: Using Ontologies to Build Virtual Worlds for the Web. In: Proc. of the IADIS International Conference WWW/Internet 2004 (ICWI 2004), Madrid, Spain, pp. 683–689. IADIS Press (2004)
7. De Troyer, O., Bille, W., Romero, R., Stuer, P.: On Generating Virtual Worlds from Domain Ontologies. In: Proc. of the 9th International Conference on Multi-Media Modeling, pp. 279–294 (2003)
8. De Troyer, O., Kleinermann, F., Pellens, B., Bille, W.: Conceptual Modeling for Virtual Reality. In: Tutorials, Posters, Panels, and Industrial Contributions of the 26th international Conference on Conceptual Modeling (ER 2007), CRPIT, Auckland, New Zealand, pp. 5–20 (2007)
9. Kleinermann, F., De Troyer, O., Mansouri, H., Romero, R., Pellens, B., Bille, W.: Designing Semantic Virtual Reality Applications. In: Proc. of the 2nd INTUITION International Workshop, pp. 5–10 (2005)
10. Vince, J.: Introduction to Virtual Reality. Springer, Heidelberg (2004)
11. Burdea, G., Coiffet, P.: Virtual Reality Technology. John Wiley & Sons, Chichester (2003)
12. Activeworlds, http://www.activewords.com
13. Quake, http://www.idsoftware.com/gmaes/quake/quake
14. Unreal, http://www.unreal.com/
15. Hartman, J., Wernecke, J.: The VRML 2.0 Handbook. Addison Wesley, London (1998)
16. Web 3D Consortium, http://www.web3d.org/x3d/specifications/
17. Selman, D.: Java3D Programming. Manning (2002)
18. Octaga, http://www.octaga.com
19. Vivaty, http://www.vivaty.com/
20. Kelly, L.: Murdock, 3ds max 5 bible. Wiley Publishing, Chichester (2003)
21. Derakhshami, D.: Introducing Maya 2008. AutoDesk Maya Press (2008)
22. Milkshape, http://chumbalum.swissquake.ch
23. AC3D, http://www.invis.com/
24. Roosendaal, T., Selleri, S.: The official Blender 2.3 Guide: Free 3D creation suite for Modeling, Animation and rendering. No Starch Press (2005)
25. Performer, http://www.sgi.com/products/software/performer
26. X3D Toolkit, http://artis.imag.fr/Software/x3D/
27. Xj3D, http://www.xj3d.org
28. Chen, P.: The Entity-Relationship Model: Towards a Unified View of Data. ACM Transactions on Database Systems 1(1), 471–522 (1981)

29. Pellens, B.: A Conceptual Modelling Approach for Behaviour in Virtual Environments using a Graphical Notation and Generative Design Patterns. Ph.D. thesis, Vrije Universiteit Brussel, Brussels, Belgium (2007)
30. Pellens, B., De Troyer, O., Bille, W., Kleinermann, F.: Conceptual modeling of object behavior in a virtual environment. In: Proceedings of Virtual Concept, pp. 93–94 (2005)
31. Pellens, B., De Troyer, O., Bille, W., Kleinermann, F., Romero, R.: An ontology-driven approach for modeling behavior in Virtual Environments. In: Meersman, R., et al. (eds.) Proceedings of Ontology Mining and Engineering and its use for Virtual Reality, pp. 1215–1224. Springer, Heidelberg (2005)
32. Pellens, B., Kleinermann, F., De Troyer, O., Bille, W.: Model-based design of virtual environment behavior. In: Zha, H., et al. (eds.) Proceedings of the 12th International Conference on Virtual Reality Systems and Multimedia, pp. 29–39. Springer, Heidelberg (2006)
33. Pellens, B., De Troyer, O., Kleinermann, F., Bille, W.: Conceptual modeling of behavior in a virtual environment. International Journal of Product and Development, Inderscience Enterprises, 14–18 (2006)
34. Pellens, B., Kleinermann, F., De Troyer, O.: Intuitively Specifying Object Dynamics in Virtual Environments using VR-WISE. In: Proc. of the ACM Symposium on Virtual Reality Software and Technology, pp. 334–337. ACM Press, New York (2006)
35. Ontoprise GmbH.: How to write F-Logic Programs covering OntoBroker version 4.3 (2006)
36. May, W.: How to Write F-Logic Programs in Florida. Institut fur Informatik. Universitat Freiburg, Germany (2006), http://dbis.informatik.uni-freiburg.de
37. ODE, http://www.ode.org/
38. PhysX, http://www.nvidia.com/object/nvidia_physx.html
39. MotionWork, http://www.motionworks.com.au/
40. Kim, G.J., Kang, K.C., Kim, H.: Software engineering of virtual worlds. In: Proceedings of the ACM symposium on Virtual Reality and Technology, pp. 131–138. ACM Press, New York (1998)
41. Wachsmuth, I., Jung, B.: Dynamic conceptualization in a mechanical-object assembly environment. Artificial Intelligence Review 10(3-4), 345–368 (1996)
42. Jung, B., Hoffhenke, M., Wachsmuth, I.: Virtual assembly with construction kits. In: Proceedings of ASME Design Engineering Technical Conferences, pp. 150–160 (1997)
43. Jung, B., Wachsmuth, I.: Integration of geometric and conceptual reasoning for interacting with virtual environments. In: Proceedings of the AAAI Spring Symposium on Multimodal Reasoning, pp. 22–27 (1998)
44. Rachuri, S., Han, Y.-H., Foufou, S., Feng, S.C., Roy, J., Wang, F., Sriram, R.D., Lyons, K.W.: A model for capturing product assembly information. Journal of Computing and Information Science in Engineering 6(1), 11–21 (2006)
45. Fenves, S.: A core product model for representing design information. Technical report NISTIR 6736, National Institute of Standards and Technology, NIST (2001)
46. Jayaram, S., Connacher, H., Lyons, K.: Virtual assembly using virtual reality techniques. Journal of Computer-Aided Design 29(8), 155–175 (1997)
47. Jayaram, S., Wang, Y., Jayaram, U., Lyons, K., Hart, P.: A virtual assembly design environment. In: Proceedings of IEEE Virtual Reality Conference, pp. 172–180 (1999)
48. Smith, G., Stuerzlinger, W.: Integration of constraints into a vr environment. In: Proceedings of the Virtual Reality International Conference, pp. 103–110 (2001)

49. Gosele, M., Stuerzlinger, W.: Semantic constraint for scene manipulation. In: Proceedings of the Spring Conference in Computer Graphics, pp. 140–146 (2002)
50. Stuerzlinger, W., Graham, S.: Efficient manipulation of object groups in virtual environments. In: Proceeding of the VR 2002 (2002)
51. Brutzman, D., Daly, L.: X3D: Extensible 3D graphics for Web Authors. Morgan Kaufmann, San Francisco (2007)
52. Arnaud, R., Brnes, M.: Collada: Sailing the gulf of 3d digital content creation. A K Peters, Ltd., Massachusetts (2006)
53. SimMechanics, http://www.mathworks.com
54. SolidWorks, http://www.solidworks.com
55. Coninx, K., De Troyer, O., Raymaekers, C., Kleinermann, F.: VR-DeMo: a Tool-supported Approach Facilitating Flexible Development of Virtual Environments using Conceptual Modelling. In: Proc. of Virtual Concept, pp. 65–80 (2006)
56. Mansouri, H., Kleinermann, F., De Troyer, O.: Detecting Inconsistencies in the Design of Virtual Environments over the Web using Domain Specific Rules. In: Proc. of the 14th International Symposium on 3D Web Technology (Web3D 2009), pp. 31–38. ACM Press, New York (2009)

Towards a General Framework for Effective Solutions to the Data Mapping Problem

George H.L. Fletcher[1],[*] and Catharine M. Wyss[2]

[1] Department of Mathematics and Computer Science
Eindhoven University of Technology, The Netherlands
g.h.l.fletcher@tue.nl
[2] School of Informatics
Indiana University, Bloomington, USA
cmw@cs.indiana.edu

Abstract. Automating the discovery of mappings between structured data sources is a long standing and important problem in data management. We discuss the rich history of the problem and the variety of technical solutions advanced in the database community over the previous four decades. Based on this discussion, we develop a basic statement of the data mapping problem and a general framework for reasoning about the design space of system solutions to the problem. We then concretely illustrate the framework with the Tupelo system for data mapping discovery, focusing on the important common case of relational data sources. Treating mapping discovery as example-driven search in a space of transformations, Tupelo generates queries encompassing the full range of structural and semantic heterogeneities encountered in relational data mapping. Hence, Tupelo is applicable in a wide range of data mapping scenarios. Finally, we present the results of extensive empirical validation, both on synthetic and real world datasets, indicating that the system is both viable and effective.

Keywords: data mapping, data integration, schema matching, schema mapping, data exchange, metadata, data heterogeneity.

1 Introduction

The emerging networked world promises new possibilities for information sharing and collaboration. These possibilities will be fostered in large part by technologies which facilitate the cooperation of autonomous data sources. Created and evolving in isolation, such data sources are maintained according to local constraints and usage. Consequently, facilitating technologies must bridge a wide variety of heterogeneities, such as differences at the system level, differences in the structuring of data, and semantic pluralism in the interpretation of data.

The world-wide-web and its myriad supporting technologies have proven very successful for overcoming the system-level heterogeneities which arise in data

---

[*] Corresponding author.

S. Spaccapietra, L. Delcambre (Eds.): Journal on Data Semantics XIV, LNCS 5880, pp. 37–73, 2009.
© Springer-Verlag Berlin Heidelberg 2009

FlightsA

Flights:

Carrier	Fee	ATL29	ORD17
AirEast	15	100	110
JetWest	16	200	220

FlightsB

Prices:

Carrier	Route	Cost	AgentFee
AirEast	ATL29	100	15
JetWest	ATL29	200	16
AirEast	ORD17	110	15
JetWest	ORD17	220	16

FlightsC

AirEast:			JetWest:		
Route	BaseCost	TotalCost	Route	BaseCost	TotalCost
ATL29	100	115	ATL29	200	216
ORD17	110	125	ORD17	220	236

Fig. 1. Three airline flight price databases, each with the same information content

sharing. However, these technologies have not addressed the difficult forms of data-level heterogeneity. At the heart of overcoming data heterogeneity is the *data mapping problem*: automating the discovery of effective mappings between autonomous data sources. The data mapping problem remains one of the longest standing issues in data management [44,56]. Data mapping is fundamental in data cleaning [10,65], data integration [48], and semantic integration [36,62]. Furthermore, mappings are the basic glue for constructing large-scale semantic web and peer-to-peer information systems [41,70]. Consequently, the data mapping problem has a wide variety of manifestations such as schema matching [6,13,26,69], schema mapping [4,31,44,81], ontology alignment [16,38,75], and model matching [56].

Fully automating the discovery of data mappings has long been recognized as a "100-year" "AI-complete" problem [24,42,52,56,74]. Consequently, solutions have typically focused on discovering simple mappings such as attribute-to-attribute schema matching [64,69]. More robust solutions to the problem must not only discover such restricted mappings, but also facilitate the discovery of the structural transformations [21,47,58,78] and complex (many-to-one-attribute) semantic functions [12,34,36,62] which inevitably arise in coordinating heterogeneous information systems [39].

Example 1. Consider three relational databases Flights A, B, and C maintaining cost information for airline routes, as shown in Fig. 1. These databases, which exhibit three different natural representations of the same information, could be managed by independent travel agencies wishing to share data. Note that mapping between these databases requires (1) matching schema elements, (2) dynamic data-metadata restructuring, and (3) complex semantic mapping. For example, mapping data from FlightsB to FlightsA involves (1) matching the Flights and Prices table names and (2) promoting data values in the Route column to attribute names. Promoting these values will dynamically create as many new attribute names as there are `Route` values in the instance of `FlightsB`. Mapping the data in `FlightsB` to `FlightsC` requires (3) a complex semantic function mapping the sum of `Cost` and `AgentFee` to the `TotalCost` column in the relations of `FlightsC`.   □

To better understand the design space of general solutions for the full data mapping problem, it is necessary to take a step back from particular instances of the problem. Such a broad perspective provides insight into the crucial aspects of the problem and into fundamental design techniques which can in turn be applied towards more robust and efficient solutions to particular data mapping scenarios.

Overview. Recognizing that data mapping is an AI-complete challenge, we study various facets of the problem with an eye towards developing a better understanding of the generic design space of data mapping solutions. In this investigation, we strive towards understanding both what data mappings are and how to go about discovering them. Our primary contributions are:

- A novel abstract definition of the data mapping problem and an application of this definition to the important special case of relational data sources (Section 2);
- a novel generic architecture for the design of effective solutions to the data mapping problem (Section 2.4); and
- an instantiation and evaluation of this architecture in the Tupelo data mapping system (Section 3), which applies an example-driven methodology for mapping discovery between relational data sources.

During the course of the paper, related research efforts are highlighted. We conclude with a discussion of research directions which build on these contributions (Section 4). The second half of this paper revises and extends [20].

2   The Data Mapping Problem

The data mapping problem has deep historical roots. We begin this Section with a brief account of the data mapping problem as it arose as a theoretical and technological problem. We then turn our attention to a formalization of this discussion, as a foundation for making practical contributions on the problem. Finally, we close the Section with an application of this formalism to the important special case of mapping between relational databases. An outcome of this discussion is a general design framework for mapping-discovery systems.

2.1   Perspectives on the Data Mapping Problem

We briefly highlight the historical roots of the data mapping problem. This account argues that data mapping is one of the oldest intellectual and practical concerns of science. We aim to show the ubiquity and generality of the problem, beyond technical motivations.

Philosophical, Linguistic, and Cognitive Roots. One could argue that the philosophical problem of *communication*, a concern since the earliest of Greek philosophers, is a manifestation of the data mapping problem. Indeed, the perplexing question of how it is that two speakers come to some common agreement during conversation can be recast as a question of how differences in perspective are resolved through mapping between world-views.

Early philosophical considerations set the stage for a wide-ranging discussion, which continues to this day, concerning the semantics, interpretation, and origins of natural and artificial languages (a nice overview of these investigations can be found in [33]). Of particular interest for background on the data mapping problem, semiotician Umberto Eco has documented the long struggle to overcome the perceived problems which stem from language and worldview heterogeneity [15]. Eco highlights early efforts on developing "universal" ontologies and artificial languages, such as those proposed in the 17th century by Dalgarno and Wilkins and more recent efforts such as the Esperanto movement and research on knowledge representation. In many ways, ongoing research efforts towards building universal knowledge bases are a continuation of this long-standing effort towards resolving, once-and-for-all, syntactic and semantic data heterogeneity [55]. Of course, outside of information systems research, investigators in linguistics and cognitive science have also focused intense sustained effort on resolving the inherent problems of mapping between heterogeneous conceptual models in biological and artificial communicative systems, e.g., [17].

Technological Roots. In the field of information systems, it was recognized early on that data mapping is a fundamental aspect of managing any large collection of data. From pioneering work on database reorganization in systems such as ExPress developed at IBM [68] in the mid-1970s, to work in the 1980s and 1990s in database schema integration [3], interoperability [54,67], and schema matching [64,69], data mapping has arisen in a wide variety of forms and guises. Moving to the late 1990's and 2000's, data mapping has resurfaced in recent work in ontology management [16,38,70,71]. In each of these areas, a key problem has been the discovery of transformations for mapping data between heterogeneous data representations. Much of this research has assumed that human users will provide these vital pieces which glue together information systems. Only recently have there been efforts to automate some aspects of the discovery of data mappings.

2.2   A Formal Presentation of the Data Mapping Problem

In this Section we give a formal generic presentation of the data mapping problem which generalizes and strives towards making more actionable the historical discussion of Section 2.1. This formalization allows us to focus on the essential aspects of the technological problem, and provides a foundation for further practical progress in the design and construction of automated data mapping solutions. There have been intense research efforts on data mapping formalisms. Recent key examples include the formalisms of Calvanese et al., Grahne & Kiricenko, and Lenzerini for data integration [9,29,48], the model management framework of Melnik et al. [56], and frameworks for ontology mapping [16,38,70,71]. Our formalism encompasses and extends its predecessors within a generalized statement of the technical *problem* of data mapping.

The Structure of Data Schemata. In data mapping, we are concerned with discovering mappings between *data schemata*, which are clearly delineated classes of structured *data objects*. For the purposes of our generic discussion in this Section, the internal structure of data objects and particular mechanisms for their construction are unimportant; therefore, we simply posit a universe $\mathbb{O}$ of data objects. A *data model* is a formalism for concretely describing the structuring of atomic data into data objects [1].

Definition 1. *A data model $\mathfrak{M}$ is a computable subset of $\mathbb{O}$.*

Examples of well-known concrete data models include the relational, XML, nested relational, and OODB data models [1].

A data(base) schema is a description of a class of data objects in terms of a particular data model [1]. For our purposes here, it is only important that a schema identifies a well-defined subset of the objects in a data model.

Definition 2. *A data schema S in a data model $\mathfrak{M}$ is a computable boolean function from $\mathfrak{M}$ to the set $\{\top, \bot\}$. A data object $D \in \mathfrak{M}$ is said to be valid with respect to S if $S(D) = \top$. We will call the set of all valid data objects with respect to S, denoted $\mathcal{D}_S = \{D \in \mathfrak{M} \mid S(D) = \top\}$, the extension of S.*

We will sometimes conflate a schema S and its extension $\mathcal{D}_S$, when it is clear from context.

The Structure of Data Mappings. We next define data mappings between schemata.

Definition 3. *A data mapping from a schema S to a schema T is a binary relation $\varphi \subseteq \mathcal{D}_S \times \mathcal{D}_T$.*

By not requiring data mappings to be *functional* relations,[1] this definition accommodates probabilistic, incomplete, and uncertain data management scenarios [11,29]. We further note that Definition 3 does not restrict us to considering *computable* data mappings. This flexibility in the formalism is likewise necessary to accommodate the wide range of possibilities for mapping scenarios. In many practical cases, however, the data mappings under consideration will not be quite so problematic.

Example 2. Consider data models

$$\mathfrak{M}_{\text{source}} = \{D_a^s, D_b^s, D_c^s, D_d^s\} \qquad \mathfrak{M}_{\text{target}} = \{D_a^t, D_b^t, D_c^t, D_d^t, D_e^t\},$$

---

[1] i.e., requiring that $\forall D \in \mathcal{D}_S$ it must be the case that $|\varphi(D)| = 1$.

schemata

$$\mathcal{D}_S = \{D_a^s, D_b^s\} \qquad\qquad \mathcal{D}_T = \{D_a^t, D_b^t, D_c^t\}$$

in $\mathfrak{M}_{\text{source}}$ and $\mathfrak{M}_{\text{target}}$, resp., and the following binary relations in $\mathfrak{M}_{\text{source}} \times \mathfrak{M}_{\text{target}}$:

$$\varphi = \{(D_a^s, D_a^t), (D_b^s, D_a^t)\}$$
$$\psi = \{(D_a^s, D_a^t), (D_a^s, D_b^t), (D_a^s, D_c^t), (D_b^s, D_a^t)\}$$
$$\chi = \{(D_a^s, D_a^t)\}.$$

Then φ is a left-total functional data mapping, ψ is a non-functional data mapping, and χ is partial functional data mapping, each from $\mathcal{D}_S$ to $\mathcal{D}_T$.   □

This example illustrates the special case of *finite* (and hence computable) data mappings. If mappings are to be specified by a human expert, then this will indeed be the case. Recently, a theoretical analysis has been undertaken for this important scenario [18].

In mapping discovery, we are ultimately interested in specifying data mappings in some concrete syntax; we capture this as follows.

Definition 4. *Let $\mathfrak{M}_{source}$ and $\mathfrak{M}_{target}$ be data models. A $\mathfrak{M}_{source} \triangleright \mathfrak{M}_{target}$ mapping language is a pair $\langle \mathcal{E}, \llbracket \cdot \rrbracket \rangle$, where:*

- *$\mathcal{E}$ is a computable set of finite strings over a finite alphabet, and*
- *$\llbracket \cdot \rrbracket$ is a computable function which maps each element of $\mathcal{E}$ to a data mapping φ, where φ is from a schema in $\mathfrak{M}_{source}$ to a schema in $\mathfrak{M}_{target}$.*

Elements of $\mathcal{E}$ are called mapping expressions. *We will use "$E \in \mathscr{L}$" as short-hand for the statement "$E \in \mathcal{E}$ for mapping language $\mathscr{L} = \langle \mathcal{E}, \llbracket \cdot \rrbracket \rangle$."*

Intuitively, $\mathcal{E}$ is the set of expressions (i.e., finite syntactic objects) of some well-defined mapping language (e.g., the relational algebra or XPath), and $\llbracket \cdot \rrbracket$ is the semantic evaluation function for the language which defines the meaning of expressions in terms of data objects in $\mathfrak{M}_{\text{source}}$ and $\mathfrak{M}_{\text{target}}$.

Before we move on to define the general data mapping problem, it is worthwhile to make the following observations. As we saw in Section 1 and Section 2.1, data mapping is pervasive in information systems and is intimately bound up not only in technological concerns but also in social concerns, since it is human activities and interests which are ultimately facilitated by these systems. In striving to capture this, it may appear that our abstract definitions of data models and mappings become too permissive and open-ended. We argue, however, that it is worthwhile to attempt to address as much of this problem space as possible at the outset, and then move on to special cases where technological progress can be made. Of course, our interests are strictly technological; when we consider specific data mapping scenarios, this abstract structure becomes grounded in actionable data models and mapping languages, as we will see in Section 2.3.

Defining the Data Mapping Problem. We are now prepared to state the general data mapping problem (DMP).

> DMP. Let S and T be data schemata in data models $\mathfrak{M}_{\text{source}}$ and $\mathfrak{M}_{\text{target}}$, respectively; φ be a data mapping from $\mathcal{D}_S$ to $\mathcal{D}_T$; and $\mathscr{L}$ be a $\mathfrak{M}_{\text{source}} \triangleright \mathfrak{M}_{\text{target}}$ mapping language. Does there exist a mapping expression $E \in \mathscr{L}$ such that $[\![E]\!] = \varphi$?

The intuition behind this characterization of the data mapping problem is as follows: during data mapping discovery, an ideal "oracle" mapping φ is typically elicited informally from a human expert (perhaps interactively using a graphical user interface in a piece-meal, step-wise fashion) or is otherwise assumed to exist (and to be verifiable), and the task at hand is to semi-automatically discover a mapping expression E in some appropriate concrete executable mapping language $\mathscr{L}$ (such as SQL, XSLT, or probabilistic relational algebra) such that the behavior of E on data objects in $\mathcal{D}_S$ is precisely that of φ.[2]

2.3   Data Mapping in Relational Databases

Note that DMP is really a *template* for specific data mapping problems. We concern ourselves in the balance of this paper with instances of DMP where the source and target schemata, S and T, are both relational, and the mapping language $\mathscr{L}$ is a relational database *query language*.[3] In this section we present the specific details of the data mapping problem for relational data sources. This concrete presentation will follow the formalism of Section 2.2. Unlike the formal presentation, however, we will now be concerned with the internal structure of data objects. Although we focus on relational databases, we note that the discussion which follows in the balance of this paper is illustrative of any data mapping scenario where the source and target schemata are structured, and the mapping language is an appropriate database query language. For example, our general approach can be readily transferred to mapping scenarios involving XML data sources (or a mix of sources from various structured data models) and mapping languages such as XPath or XSLT.

Relational Data Model. We follow a variation of the general framework for relational data objects as presented by Wyss et al. [78] and the uniform data model of Jain et al. [37]. In short, we have that: a *tuple* is a finite set of ordered pairs of uninterpreted symbols (i.e., attribute-value pairs); a *relation* is a named finite set of tuples; and a *database* is a named finite set of uniquely named relations. The *schema* of a relation is its name taken together with the set of attribute names of its constituent tuples; the schema of a database is its name

---

[2] It is also interesting to consider an extension of DMP, where E is only required to *approximate* the behavior of φ within a given error-bound.

[3] i.e., a language which specifies mappings on schemas in the relational data model which are computable and generic partial-functions [1].

taken together with the set of schemata of its relations. All symbols (including relation and database names) are assumed to be from some enumerable domain $\mathbb{U}$ of uninterpreted atomic objects (e.g., Unicode strings, JPEG images, MP3 music files, PDF documents, etc.).

Example 3. Consider the FlightsA database from Figure 1. In the relational data model, this database has encoding $\langle \text{FlightsA}, D \rangle$ where

$$D =$$
$$\{\langle \text{Flights}, \{ \langle \text{Carrier}, \text{AirEast} \rangle, \langle \text{Fee}, 15 \rangle, \langle \text{ATL29}, 100 \rangle, \langle \text{ORD17}, 110 \rangle \} \rangle,$$
$$\langle \text{Flights}, \{ \langle \text{Carrier}, \text{JetWest} \rangle, \langle \text{Fee}, 16 \rangle, \langle \text{ATL29}, 200 \rangle, \langle \text{ORD17}, 220 \rangle \} \rangle \}.$$

$\square$

Relational Mapping Languages. Research on mapping languages (i.e., the set $\mathscr{L}$ in the DMP definition) for the relational data model has been going strong for over 30 years. At the core of almost all of these languages is the relational algebra (RA). In what follows, we assume familiarity with RA.

Example 4. Recall database $\langle \text{FlightsA}, D \rangle$ from Example 3. Suppose we wish to extract Carrier values from this database and place the output in a relation named Companies. The following RA query does the trick:

$$\rho_{\text{Flights} \to \text{Companies}}^{\text{rel}}(\pi_{\text{Carrier}}(\langle \text{FlightsA}, D \rangle)) = \langle \text{FlightsA}, D' \rangle$$

where the superscript rel on the rename operator ρ indicates relation renaming, and

$$D' =$$
$$\{\langle \text{Companies}, \{\langle \text{Carrier}, \text{AirEast} \rangle\} \rangle, \langle \text{Companies}, \{\langle \text{Carrier}, \text{JetWest} \rangle\} \rangle\}.$$

$\square$

For an overview of the rich variety of relational mapping languages, see [1].

Relational Data Mapping Problem. With the relational data model and RA as an example relational mapping language in hand, we are now in a position to turn to a concrete presentation of RelationalDMP, the DMP for relational data sources.

> RelationalDMP. Let S and T be relational data schemata, φ be a data mapping from $\mathcal{D}_S$ to $\mathcal{D}_T$, and $\mathscr{L}$ be a relational query language. Does there exist an expression $E \in \mathscr{L}$ such that $[\![E]\!] = \varphi$?

For the balance of this paper, we will be concerned with investigating this important subclass of DMP problems.

2.4   A Framework for Data Mapping Systems

We now turn to a general overview of the design and construction of solutions for instances of RelationalDMP. Our design is driven by the following crucial observation, which encapsulates a long-running analysis of the various subproblems of RelationalDMP in the data mapping literature (e.g., [36,39,43,47,56,58]):

> *Data mapping discovery for structured data sources consists of two distinct principle tasks: discovery of semantic functions and discovery of structural mapping queries.*

Semantic functions operate at the token level (i.e., operate on tokens in $\mathbb{U}$), mapping data values between data sources. These functions interpret the internal structure of atoms in $\mathbb{U}$, and hence rely on information which is external to the information systems of which a data mapping system is a component. This information can be codified (for example) in an ontology, but it is important to note that our work does not presuppose pre-existing ontologies, nor even a shared vocabulary of tokens.

In RelationalDMP, *mapping queries* operate at the structural level, corresponding to traditional structural database transformations between database schemas. Discovering semantic functions and discovering mapping queries both require unsupervised learning from data instances and/or supervised learning using domain knowledge. For these tasks it is possible to leverage the large body of techniques which have been developed over the last century in the Machine Learning, Artificial Intelligence, and Data Mining communities [61].

We propose a generic architecture for RelationalDMP solutions which reflects our design observation, illustrated in Figure 2. Input to the mapping discovery process includes, at the very least, source/target database schemas and instances. If available, the discovery process can also use domain knowledge elicited from external sources (e.g., human input, system logs, etc.). This architecture, which we now outline, cleanly captures the division of labor in data mapping implied by the design thesis.

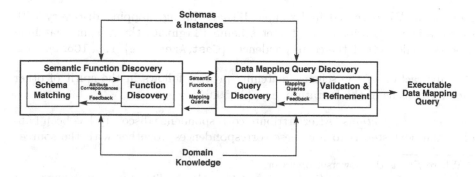

Fig. 2. Generic architecture for discovery of executable data mapping queries

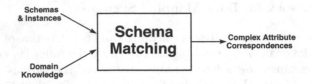

Fig. 3. Complex attribute correspondence discovery

Discovering Semantic Functions. The first step in data mapping is to discover semantic functions (Figure 2). This process involves (1) *schema matching* – identifying the correspondences between those attributes of the source and target schemas that are semantically related – and then (2) discovering the semantic functions which provide the actual mappings between the data values of these corresponding attributes. Note that this may be an iterative process, with information gained during function discovery used in further schema matching (Figure 2).

Discovery of Complex Attribute Correspondences. Schema matching takes as input the source and target schemas, S and T, and instances if available. If available, domain knowledge can also be used to supervise the discovery of attribute correspondences (Figure 3). Together, these inputs serve to guide the selection of good pairings between the schema elements of the source S and the target T. This process can be abstractly presented as follows. The relationship between S and T is encapsulated as a boolean function M on the set $\mathcal{P}(\text{schema}(S)) \times \text{schema}(T)$, and is typically defined externally to the schema matching mechanism itself.[4] The output of this component is a set of complex (i.e., many-one) correspondences [22,64] between the attributes of S and those of T that satisfy M:[5]

$$\{(\overline{a}_{\text{source}}, a_{\text{target}}) \mid \overline{a}_{\text{source}} \subseteq \text{schema}(S), a_{\text{target}} \in \text{schema}(T),$$
$$\& \; M(\overline{a}_{\text{source}}, a_{\text{target}}) = \top\}.$$

Example 5. We observed in Example 1 that, during mapping discovery with source schema `FlightsB` and target schema `FlightsC`, the schema matching process would output the correspondence $(\{\text{Cost}, \text{AgentFee}\}, \text{TotalCost})$. □

Our abstraction of the schema matching process follows in the spirit of other such formalisms in the literature, e.g., [23,56,64].

Discovery of Functions. After attribute correspondence discovery has been handled, the next step is to use these correspondences, together with the source/

---

[4] Where $\mathcal{P}(\cdot)$ is the powerset operator.

[5] The case of many-many (i.e., m-n) matchings in the literature [64] reduces to a special case of many-one matchings, namely, where one is interested in a set of n many-one correspondences.

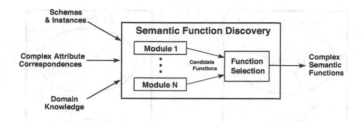

Fig. 4. Semantic function discovery

target schemas and instances, to discover the semantic functions which map between corresponding source and target data values (Figure 4). As with schema matching, domain knowledge can also be used to supervise the discovery process, if it is available. It is generally recognized that a *modular* approach to function discovery (Figure 4) is necessary to accommodate the wide variety of possibilities for token transformation scenarios [12]. Hence, our design architecture indicates specialized modules responsible for discovering specific classes of functions (e.g., string functions [12,73], date/time conversions, real-time currency exchange, image conversion, etc.). The last step then is to select the final functions from among the candidates suggested by these modules (Figure 4).

We can abstract the process of function discovery as follows. Concrete semantic functions map sets of atoms $\bar{s}$ in a source token space (i.e., a subset of $\mathbb{U}$) to atoms t in a target token space (i.e., another subset of $\mathbb{U}$). These token spaces reside in source and target contexts of interpretation, respectively (Figure 5). In these contexts of interpretation, bijective "meaning" functions m_{source} and m_{target} associate these tokens with some objects $\overline{O}_s$ and O_t, respectively, in domains of discourse which are of interest to the users of the source and target information systems, respectively:

$$m_{\text{source}}(\bar{s}) = \overline{O}_s \qquad m_{\text{target}}(t) = O_t$$

Analogous to the match function M in schema matching, the relationships between objects in the source and target domains of discourse are encapsulated in a discourse-mapping function f^* (Figure 5), whose derivation is external to the function discovery process. Now, given a complex attribute correspondence $(\bar{a}_{\text{source}}, a_{\text{target}})$, the goal of semantic function discovery is to find a concrete function f (Figure 5) which maps,[6] on a per-tuple basis, any instance $\bar{s}$ of attributes $\bar{a}_{\text{source}}$ in the source token space to an instance t of attribute a_{target} in the target token space such that

$$f(\bar{s}) = m_{\text{target}}(f^*(m_{\text{source}}(\bar{s}))) = t.$$

In other words, f abides by the semantics of both the source and target schemata in their contexts of use.

---

[6] For further discussion of the "semantics" of applying semantic functions, please see Section 3.5 below.

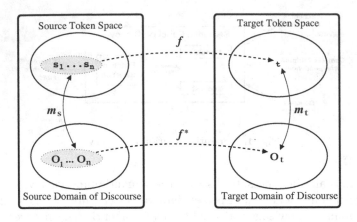

Fig. 5. Semantic mappings between contexts-of-interpretation

Example 6. As we saw in Example 5, the attributes Cost and AgentFee in database FlightsB correspond to the attribute TotalCost in the relations of database FlightsC. Once this correspondence has been determined, for mapping instances of FlightsB to instances of FlightsC, we require a semantic function f for this correspondence which interprets the tokens of all three attributes appropriately as numbers in the reals[7] and transforms them as follows:

$$f : \text{Cost} + \text{AgentFee} \longmapsto \text{TotalCost}$$

applied to each tuple in the Prices relation of FlightsB.   □

While solutions abound in the literature for one-to-one schema matching [64], the database community has only recently begun to make strong progress on the issues of *complex* (i.e., many-to-one) schema matching and semantic function discovery. Primarily, supervised approaches (i.e., using domain knowledge elicited via GUIs, etc.) have been explored in the literature [8,12,14,23,30,34,41,62,81]. The design and analysis of approaches to complex schema matching and semantic function discovery continues to be an extremely important area of investigation in data mapping.

Discovering Data Mapping Queries. After determining appropriate semantic functions, the second critical step in data mapping is to discover executable queries (Figure 2). These queries perform restructuring of data objects and apply the previously discovered semantic functions. Note that this process may be iterative. In the context of structural transformations which dynamically modify the input schemata, further rounds between semantic function discovery and query discovery may be necessary (Figure 2).

---

[7] Or, if appropriate, as currency values, applying exchange rates as necessary if multiple currencies are involved, etc.

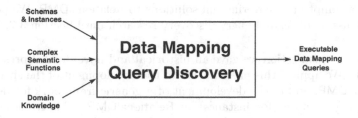

Fig. 6. Data mapping query discovery

Example 7. Continuing Example 6, we recall from the discussion of Example 1 that structural mapping from the `FlightsB` schema to the `FlightsC` schema requires data-metadata transformation in a language such as the federated interoperable relational algebra, a variation of which we develop in Section 3.1 below. As we will see, this language is a natural extension of the RA that includes data-metadata structural transformations. In this algebra, the following mapping query restructures data under the `FlightsB` schema to conform with the `FlightsC` schema:

$$\mu_{\text{Carrier, Cost}}\left(\wp_{\text{Carrier}}\left(\rho_{\text{Cost}\rightarrow\text{BaseCost}}\left(\lambda_{f, \text{ Cost, AgentFee}}^{\text{TotalCost}}(\text{FlightsB})\right)\right)\right).$$

Note the λ operator for application of semantic functions. In this case, the function f from Example 6 is applied to the values in the attributes `Cost` and `AgentFee`, and the results are placed in target attribute `TotalCost`.   □

Source/target schemata, S and T, instances of these schemata (if available), the complex semantic functions discovered between these schemata in the previous step, and domain knowledge (if explicitly available) are given as input to the query discovery component of the mapping framework (Figure 6). Collectively, the constraints of these inputs delineate an ideal data mapping φ from S to T. The output of this process is a solution for the RelationalDMP instance $\langle S, T, \mathscr{L}, \varphi \rangle$, where $\mathscr{L}$ is an appropriate query language for S, T, and φ. In other words, the output of this module is an executable data mapping expression E (or a set of candidate mappings) in some concrete query language $\mathscr{L}$ that transforms any valid instance of schema S into a corresponding valid instance of schema T (i.e., respecting φ). The final sub-step of mapping discovery is validation and refinement of the discovered mapping(s) (Figure 2).

Query discovery is the least explored aspect of RelationalDMP(and of the general DMP). Only a handful of systems attack aspects of query discovery [4,12,31,32,40,57,59,60,66,81], and hence the problem is not very clearly recognized in the literature. Since it *is* generally recognized in the literature that executable data mapping queries are the crucial glue in the various manifestations of data mapping discussed in Section 1, clearly continued efforts on understanding DMP are warranted. We note here that to the best of our knowledge, our research in this Section and Section 3 below on building a mapping query discovery solution (i.e., an instantiation of the framework presented in this Section) is the first attempt at discovering full solutions to RelationalDMP. We postpone a fuller discussion of related query discovery research until Section 3.7, below.

Remarks. In this Section we gave an historical and novel formal presentation of the DMP. We applied this presentation to the development of the special case of RelationalDMP, and to the development of a generic framework for designing data mapping solutions for instances of RelationalDMP. This framework was based on the observation that such systems should clearly separate the discovery of transformations for semantic heterogeneity from the discovery of queries for structural heterogeneity. We next turn to the development of a system which instantiates the insights of this discussion.

3   Data Mapping as Search

We next present the Tupelo data mapping system for semi-automating the discovery of executable[8] data mapping expressions between heterogeneous relational data sources. Tupelo is an example driven system, generating mapping expressions for interoperation of heterogeneous information systems which involve schema matching, dynamic data-metadata restructuring, and complex (many-to-one) semantic functions. For example, Tupelo can generate the expressions for mapping between instances of the three airline databases in Figure 1. The design of Tupelo is guided by the generic framework for RelationalDMP solutions developed in Section 2.4.

Previous solutions have not clearly separated each of the subproblems associated with data mapping discovery – mixing, merging, and/or conflating various aspects of semantic function discovery and query discovery. This has lead to a somewhat opaque research literature with inconsistent terminology and duplication of effort. The development of Tupelo clarifies, complements, and extends the existing approaches in the literature. In particular, Tupelo is the first data mapping system to

- propose and validate the mapping query discovery process as an example-driven search problem;
- explicitly modularize the various aspects of data mapping query discovery;
- seamlessly incorporate complex semantic functions in a complete, executable mapping language; and
- generate mapping queries which incorporate the full range of data-metadata structural transformations necessary to overcome heterogeneity in relational data sources.

Data mapping in Tupelo is built on the novel perspective of mapping discovery as an example driven search problem. We develop the Tupelo mapping language $\mathscr{L}$ in Section 3.1 and the Rosetta Stone principle behind this example-driven approach in Section 3.2. We then discuss how Tupelo leverages Artificial Intelligence

---

[8] By *executable*, we mean that the discovered mapping queries must be in a concrete mapping language such as SQL or RA (cf., Sections 2.2-2.3, above).

(AI) search techniques to generate mapping expressions (Sections 3.3 and 3.4). After this, we discuss how generic query languages such as $\mathscr{L}$ can be extended naturally in this setting to accommodate complex semantic functions which have previously discovered (Section 3.5). We then present experimental validation of the system on a variety of synthetic and real world scenarios (Section 3.6) which indicates that the Tupelo approach to data mapping is both viable and effective. We conclude the Section with a discussion of related research (Section 3.7).

3.1 Dynamic Relational Data Mapping with Tupelo

Recall from Figure 2 and the discussion of Section 2.3 that a critical component of data mapping is the discovery of executable data mapping queries. Tupelo generates an effective mapping from a source relational schema S to a target relational schema T, under the assumption that semantic function discovery has been successfully completed. The system discovers this mapping using (1) example instances s of S and t of T and (2) illustrations of any complex semantic mappings between the schemas. Mapping discovery in Tupelo is a completely syntactic and structurally driven process which does not make use of a global schema or any explicit domain knowledge beyond that encapsulated in the input semantic functions [6].

The mapping language $\mathscr{L}$ used in Tupelo provides support for both simple schema matching and richer structural transformations.

FIRA. Recently, Wyss and colleagues have developed a relational language framework for *metadata integration* [77,78,79]. This framework consists of a federated relational data model, a variation of which was introduced in Section 2.3, and two equivalent relational query languages: the Federated Interoperable Relational Algebra (FIRA) and the Federated Interoperable Structured Query Language (FISQL). These languages, FIRA/FISQL, (1) are principled extensions to relational algebra/SQL (resp.) that include metadata querying and restructuring capabilities; and (2) generalize the notion of relational transpose, providing a notion of *transformational completeness* for relational metadata [78]. Applications of the FIRA/FISQL framework include OLAP, schema browsing, and real-time interoperability of relational sources in federated information systems [78]. True data integration presupposes metadata integration, and FIRA/FISQL contributes to the study of query languages specifically by advancing the understanding of languages that offer robust metadata integration capabilities.

$\mathscr{L}$: Tupelo's Take on FIRA. Tupelo generates expressions in a fragment $\mathscr{L}$ of FIRA. The operators in this fragment extend the RA (Section 2.3) with dynamic structural transformations [47,65,78]. These include operators for dynamically promoting data to attribute and relation names (i.e., to "metadata"), a simple merge operator [77], and an operator for demoting metadata to data values. The $\mathscr{L}$ operations are intuitively summarized in Table 1. A more detailed discussion and comparison of full FIRA to the wealth of alternative relational query languages can be found in [78].

Table 1. $\mathscr{L}$ operators for dynamic relational data mapping

Operation	Effect
$\pi, \cup, \times$	*Regular relational operations.*
$\rho_{A \to B}^{att/rel}(R)$	*Rename* attribute/relation A to B in relation R.
$\uparrow_A^B (R)$	*Promote* attribute A to metadata. $\forall t \in R$, append a new active attribute named $t[A]$ with value $t[B]$.
$\to_A^B (R)$	*Dereference* attribute A on B. $\forall t \in R$, append a new active attribute named B with value $t[t[A]]$.
$\downarrow_{A,B} (R)$	*Demote* metadata. Cartesian product of relation R with a binary relation (with schema $\{A, B\}$) containing the metadata of R.
$\wp_A(R)$	*Partition* on attribute A. $\forall V \in \pi_A(R)$, output a new relation named V, where $t \in V$ iff $t \in R$ and $t[A] = V$.
$\Sigma(\mathcal{D})$	*Generalized union* of database $\mathcal{D}$. Outputs an unnamed outer union of all relations in $\mathcal{D}$.
$\bot_A(R)$	*Drop* attribute A from relation R.
$\mu_{\bar{A}}(R)$	*Merge* tuples in R based on compatible values in attributes $\bar{A}$.

Example 8. Consider the transformation of instances from `FlightsB` to instances of `FlightsA` in Figure 1. This mapping can be expressed in $\mathscr{L}$ as

$$\rho_{\text{AgentFee} \to \text{Fee}}^{att}\left(\rho_{\text{Prices} \to \text{Flights}}^{rel}\left(\mu_{\text{Carrier}}\left(\bot_{\text{Route}}\left(\bot_{\text{Cost}}\left(\uparrow_{\text{Route}}^{\text{Cost}}(\text{FlightsB})\right)\right)\right)\right)\right)$$

which breaks down as follows:

$R_1 := \uparrow_{\text{Route}}^{\text{Cost}} (\text{FlightsB})$
> Promote `Route` values to attribute names with corresponding `Cost` values.

$R_2 := \bot_{\text{Route}}(\bot_{\text{Cost}}(R_1))$
> Drop attributes `Route` and `Cost`.

$R_3 := \mu_{\text{Carrier}}(R_2)$
> Merge tuples on `Carrier` values.

$R_4 := \rho_{\text{AgentFee} \to \text{Fee}}^{att}(\rho_{\text{Prices} \to \text{Flights}}^{rel}(R_3))$
> Rename attribute `AgentFee` to `Fee` and relation `Prices` to `Flights`.

The output relation R_4 is exactly `FlightsA`.   □

The original FIRA algebra is complete for the full data-metadata mapping space for relational data sources [78]. The fragment we use in Tupelo maintains the full data-metadata restructuring power of this language. The operators in our

$\mathscr{L}$ focus on bulk structural transformations (via the $\rightarrow$, $\uparrow$, $\downarrow$, $\wp$, $\times$, Ψ, and μ operators) and schema matching (via the rename operator ρ). We view application of selections (σ) as a post-processing step to filter mapping results according to external criteria, since it is known that generalizing selection conditions is a nontrivial problem [41]. Hence, Tupelo does not consider applications of the relational σ operator. Note that using a language such as $\mathscr{L}$ for data mapping blurs the distinction between schema matching (i.e., finding appropriate renamings via ρ) [64] and schema mapping [59] since $\mathscr{L}$ encompasses these basic mapping disciplines. It is for this reason that we refer to RelationalDMP (and in general, DMP) as a *data* mapping problem.

3.2   The Rosetta Stone Principle

An integral component of the Tupelo system is the notion of "critical" instances $\mathfrak{s}$ and $\mathfrak{t}$ which succinctly characterize the structure of the source and target schemas S and T, respectively. These instances illustrate the same information structured under both schemas. The *Rosetta Stone Principle* states that such critical instances can be used to drive the search for data mappings in the space of transformations delineated by the operators in $\mathscr{L}$ on the source instance $\mathfrak{s}$. Guided by this principle, Tupelo takes as input critical source and target instances which illustrate all of the appropriate restructurings between the source and target schemas.

Example 9. The instances of the three airline databases presented in Figure 1 illustrate the same information under each of the three schemas, and are examples of succinct critical instances sufficient for data mapping discovery between the FlightsA, FlightsB, and FlightsC databases.   □

Critical Instance Input and Encoding. Critical instances can be easily elicited from a user via a visual interface akin to the Lixto data extraction system [28] or visual interfaces developed for interactive schema mapping [4,8,59,72]. In Tupelo, critical instances are articulated by a user via a front-end graphical user interface which has been developed for the system (Figure 7). Since critical instances essentially illustrate one entity under different schemas, we also envision that much of the process of generating critical instances can be semi-automated using techniques developed for entity/duplicate identification and record linkage [6,76].

Critical instances are encoded internally in *Tuple Normal Form* (TNF). This normal form, which encodes databases in single tables of fixed schema, was introduced by Litwin et al. as a standardized data format for database interoperability [53]. More recently, this flexible normal form for data has been successfully used in a variety of investigations and systems, e.g., [2,80]. Tupelo makes full use of this normal form as an internal data representation format. Given a relation R, the TNF of R is computed by first assigning each tuple in R a unique ID and then building a four column relation with attributes TID, REL, ATT, VALUE,

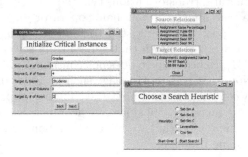

Fig. 7. Tupelo graphical user interface

corresponding to tuple ID, relation name, attribute name, and attribute value, respectively. The table is populated by placing each tuple in R into the new table in a piecemeal fashion. The TNF of a database is the single table consisting of the union of the TNF of each relation in the database.

Definition 5. *Let D be a database with name* d. *Given a tuple*

$$t = \{\langle a_1, v_1 \rangle, \ldots, \langle a_n, v_n \rangle\}$$

in relation r *of D, let $\mathring{t}$ denote the relation of n tuples:*

$$\mathring{t} = \big\langle \mathsf{d}, \{\langle \mathsf{TID}, t \rangle, \langle \mathsf{REL}, r \rangle, \langle \mathsf{ATT}, a_1 \rangle, \langle \mathsf{VALUE}, v_1 \rangle,$$
$$\ldots, \langle \mathsf{TID}, t \rangle, \langle \mathsf{REL}, r \rangle, \langle \mathsf{ATT}, a_n \rangle, \langle \mathsf{VALUE}, v_1 \rangle\} \big\rangle$$

where $t = f(t)$, for some injection f of D into $\mathbb{U}$.[9] Then, the tuple normal form *of D is the database*

$$\mathsf{TNF}(D) = \bigcup_{R \in D} \bigcup_{t \in R} \mathring{t}$$

containing a single relation named d.

Note that $\mathsf{TNF}(D)$ is well-defined (i.e., unique up to TID values). We will often blur the fact that $\mathsf{TNF}(D)$ is a database and treat it simply as a solitary unnamed relation.

Example 10. We illustrate TNF with the encoding of the instance of database `FlightsC` from Figure 1:

---

[9] i.e., t is a fresh symbol uniquely identifying tuple t.

TID	REL	ATT	VALUE
t_1	AirEast	Route	ATL29
t_1	AirEast	BaseCost	100
t_1	AirEast	TotalCost	115
t_2	AirEast	Route	ORD17
t_2	AirEast	BaseCost	110
t_2	AirEast	TotalCost	125
t_3	JetWest	Route	ATL29
t_3	JetWest	BaseCost	200
t_3	JetWest	TotalCost	216
t_4	JetWest	Route	ORD17
t_4	JetWest	BaseCost	220
t_4	JetWest	TotalCost	236

□

The TNF of a database can be built in SQL using the "system tables" of a DBMS [2,80]. The benefits of normalizing input instances in this manner with a fixed schema include (1) ease and uniformity of handling of the data; (2) both metadata and data can be handled directly in SQL; and (3) sets of relations are encoded as single tables, allowing natural multi-relational data mapping from databases to databases with the use of conventional technologies.

3.3   Data Mapping as a Search Problem

In Tupelo the data mapping problem is seen fundamentally as a search problem. Given Rosetta Stone critical instances s and t of the source and target schemas, data mapping is resolved as an exploration of the transformation space of $\mathscr{L}$ on the source instance s. Search successfully terminates when the target instance t is located in this space. Upon success, the transformation path from the source to the target is returned.[10] This search process is illustrated in Figure 8. In this Section we describe this process in more detail.

Search Algorithms. We work in the classic problem-space model [45]. In particular, a *problem space* is a pair $\langle \mathbb{S}, \mathbb{F} \rangle$, where $\mathbb{S}$ is a set of *states*, and $\mathbb{F}$ is a set of *state transition* partial functions on $\mathbb{S}$. A *search problem* consists of a problem space $\langle \mathbb{S}, \mathbb{F} \rangle$, a designated *start state* $s \in \mathbb{S}$, and a *goal test* function goal which maps states in $\mathbb{S}$ to $\{\top, \bot\}$.[11] A *solution* to a search problem is a sequence of transitions $\tau_1, \ldots, \tau_n \in \mathbb{F}$ such that $\mathsf{goal}(\tau_n(\cdots \tau_1(s))) = \top$. In terms of this model, Tupelo takes as input to the search process: Rosetta Stone source instance s and target instance t, the set of $\mathscr{L}$ transformations, and a goal test

---

[10] Note that there may be more than one path from s to t; we just return the shortest solution path (i.e., smallest mapping expression). Although the current implementation does not do so, it is straightforward to extend Tupelo to (1) present a discovered solution to the user; (2) allow the user to *deny* or *confirm* the solution; and (3) if denied, continue the search for an acceptable solution. Such an adaptation is outside of the scope of this investigation.

[11] Note that in practice, the search space $\mathbb{S}$ is not explicitly represented, but rather is implicit in the start-state and set of transition functions.

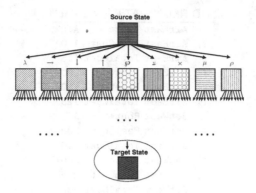

Fig. 8. Search space for data mapping discovery

which checks if a given state n is a superset of t (i.e., if t is derivable from n by filtering out tuples).

There are two general methodologies for discovering solutions to search problems: *uninformed* search and *informed* search. Both are systematic approaches to discovering a path of transformations in a search space, differing primarily in their use of external knowledge not explicit in the graph structure of the space. In particular, uninformed methods are brute-force approaches for traversing a search space, and informed methods use some external hints, in the form of *heuristics*, to guide the process. At any point during search, a choice needs to be made in selecting the next search node to visit. For each neighbor n of the current state, we can assign a *cost* value via an *evaluation function*, estimating the cost to get from the start state s, through n, to a target state:

$$\mathsf{eval}(n) = \mathsf{past}(n) + \mathsf{future}(n)$$

where $\mathsf{past}(n)$ is the known cost of reaching n from s, and $\mathsf{future}(n)$ is an estimate of the cost of getting from n to a goal state. A node with lowest cost amongst unexplored nodes is selected next for exploration. Uninformed methods ignore (or rather, do not have access to) $\mathsf{future}(n)$ during this calculation of cost. Informed methods make full use of $\mathsf{future}(n)$ in determining $\mathsf{eval}(n)$ (and may or may not ignore $\mathsf{past}(n)$). The classic "Best-First" search algorithm can be specialized to the standard uniformed Breadth-First search and informed A^* search (described below) algorithms in this manner [61].

In many domains, it is difficult to construct useful heuristic $\mathsf{future}(\cdot)$ functions [61]. In such cases, one is often limited to variations of brute-force search. The branching factor of the data mapping search space is proportional to $|s| + |t|$ and hence quite high, ruling out the use of such search methods. Fortunately, data mapping is a domain where it *is* possible to develop useful search heuristics (and would be impractical if this were not the case), and using them for intelligent exploration of the search space greatly reduces the number of states visited. Hence, we focus on informed search methodologies in the balance of our discussion. We return to the issue of developing heuristics for data mapping in Section 3.4.

Informed Search in Tupelo. Due to their simplicity and effectiveness, we chose to implement the heuristic based A^*, *Iterative Deepening A^** (IDA), and *Recursive Best-First Search* (RBFS) search algorithms from the AI literature [45,46,61]. In the exploration of a state space, these algorithms use a heuristic function to rank states and selectively search the space based on the rankings. The evaluation function eval() for ranking a search state $\mathfrak{n}$ is calculated as above, with past($\mathfrak{n}$) equal to the number of transitions to reach $\mathfrak{n}$ from $\mathfrak{s}$, and where future($\mathfrak{n}$) = $h(\mathfrak{n})$, for some heuristic function h which makes an "educated guess" of the distance of $\mathfrak{n}$ from the target state. Search begins at the source critical instance $\mathfrak{s}$ and continues until the current search state is a structurally identical superset of the target critical instance $\mathfrak{t}$ (i.e., the current state contains $\mathfrak{t}$). The transformation path from $\mathfrak{s}$ to $\mathfrak{t}$ gives a basic mapping expression in $\mathscr{L}$. After this expression has been discovered, filtering operations (via relational selections σ) must be applied if necessary according to external criteria, as discussed in Section 3.1. The final output of Tupelo is an expression for mapping instances of the source schema to corresponding instances of the target schema.

A^* is a special case of the general best-first search strategy [61]. A^* search is just best-first search called with an eval() function such that the future() component never *overestimates* the distance to a goal state. We used A^* search to develop search heuristics in early implementations of Tupelo. Unfortunately, the cost of maintaining the search queues quickly becomes impractical (given an exponential search space). Hence we were driven to explore memory-limited alternatives to best-first search.

The two search algorithms finally used in Tupelo, IDA and RBFS, operate under more practical conditions. In particular, each of these algorithms uses memory linear in the depth of search; although they both perform redundant explorations (compared to best-first search), they do not suffer from the exponential memory use of basic A^* best-first search which led to the ineffectiveness of early implementations of Tupelo. Furthermore, they both achieve performance asymptotically equivalent to A^*, as most of the work is done on the final search-frontier during a successful search.[12] In a nut-shell, these algorithms operate as follows:

- IDA performs a depth-bounded depth-first search of the state space using the eval()-rankings of states as the depth bound, iteratively increasing this bound until the target state is reached [45].
- RBFS performs a localized, recursive best-first exploration of the state space, keeping track of a locally optimal eval()-value and backtracking if this value is exceeded [46].

These two simple algorithms proved to be effective in the discovery of mapping expressions. To further improve performance of the search algorithms, we also

---

[12] In fact, they may even run faster than A^* in some cases due to lower memory management overhead [45].

employed the simple rule of thumb that "obviously inapplicable" transformations should be disregarded during search. For example if the current search state has all attribute names occurring in the target state, there is no need to explore applications of the attribute renaming operator. We incorporated several such simple rules in Tupelo.

3.4   Search Heuristics

Heuristics are used to intelligently explore a search space, as discussed in Section 3.3. A search heuristic $h(\mathfrak{n})$ estimates the distance, in terms of number of intermediate search states, of a given database $\mathfrak{n}$ from the target database $\mathfrak{t}$. A variety of heuristics were implemented and evaluated. This section briefly describes each heuristic used in Tupelo.

Set Based Similarity Heuristics. Three simple heuristics measure the overlap of values in database states. Heuristic h_1 measures the number of relation, column, and data values in the target state which are missing in state $\mathfrak{n}$:

$$h_1(\mathfrak{n}) = \quad |\pi_{\mathsf{REL}}(\mathfrak{t}) - \pi_{\mathsf{REL}}(\mathfrak{n})|$$
$$+ |\pi_{\mathsf{ATT}}(\mathfrak{t}) - \pi_{\mathsf{ATT}}(\mathfrak{n})|$$
$$+ |\pi_{\mathsf{VALUE}}(\mathfrak{t}) - \pi_{\mathsf{VALUE}}(\mathfrak{n})|.$$

Here, π is relational projection on the TNF of $\mathfrak{n}$ and $\mathfrak{t}$, and $|R|$ denotes the cardinality (i.e., number of tuples) of relation R. Heuristic h_2 measures the minimum number of data promotions ($\uparrow$) and metadata demotions ($\downarrow$) needed to transform $\mathfrak{n}$ into the target $\mathfrak{t}$:

$$h_2(\mathfrak{n}) = \quad |\pi_{\mathsf{REL}}(\mathfrak{t}) \cap \pi_{\mathsf{ATT}}(\mathfrak{n})|$$
$$+ |\pi_{\mathsf{REL}}(\mathfrak{t}) \cap \pi_{\mathsf{VALUE}}(\mathfrak{n})|$$
$$+ |\pi_{\mathsf{ATT}}(\mathfrak{t}) \cap \pi_{\mathsf{REL}}(\mathfrak{n})|$$
$$+ |\pi_{\mathsf{ATT}}(\mathfrak{t}) \cap \pi_{\mathsf{VALUE}}(\mathfrak{n})|$$
$$+ |\pi_{\mathsf{VALUE}}(\mathfrak{t}) \cap \pi_{\mathsf{REL}}(\mathfrak{n})|$$
$$+ |\pi_{\mathsf{VALUE}}(\mathfrak{t}) \cap \pi_{\mathsf{ATT}}(\mathfrak{n})|.$$

Heuristic h_3 takes the maximum of h_1 and h_2 on $\mathfrak{n}$:

$$h_3(\mathfrak{n}) = max\{h_1(\mathfrak{n}), h_2(\mathfrak{n})\}.$$

Databases as Strings: The Levenshtein Heuristic. Viewing a database as a *string* leads to another heuristic. Suppose $\mathfrak{x}$ is a database in TNF with m tuples

$$\langle t_1, r_1, a_1, v_1 \rangle, \ldots, \langle t_n, r_m, a_m, v_m \rangle.$$

For each such tuple, let $s_i = r_i \star a_i \star v_i$, where $\star$ is string concatenation. Define $\mathtt{string}(\mathfrak{x})$ to be the string $s_1 \star \cdots \star s_m$, where $s_1, \ldots, s_m$ is a lexicographic ordering of the m strings s_i, potentially with repetitions.

Example 11. Recall the TNF of database `FlightsC` from Example 10:

⟨t1 AirEast Route ATL29⟩, ⟨t1 AirEast BaseCost 100⟩,

... , ⟨t4 JetWest BaseCost 220⟩, ⟨t4 JetWest TotalCost 236⟩.

Transforming each tuple into a string and then sorting these strings, we have

string(FlightsC) =

AirEastBaseCost100AirEastRouteATL29 ··· JetWestTotalCost236.

□

The *Levenshtein distance* between string w and string v, $L(w, v)$, is defined as the least number of single character insertions, deletions, and substitutions required to transform w into v [49]. Using this metric, we define the following *normalized Levenshtein heuristic*:

$$h_L(\mathfrak{n}) = round\left(k \frac{L(\text{string}(\mathfrak{n}), \text{string}(\mathfrak{t}))}{max\{|\text{string}(\mathfrak{n})|, |\text{string}(\mathfrak{t})|\}} \right)$$

where $|w|$ is the length of string w, $k \geqslant 1$ is a scaling constant (scaling the interval $[0, 1]$ to $[0, k]$), and $round(y)$ is the integer closest to y.

Databases as Term Vectors: Euclidean Distance. Another perspective on a database is to view it as a document vector over a set of terms [5]. Let $A = \{a_1, \ldots, a_n\}$ be the set of tokens occurring in the source and target critical instances (including attribute and relation names), and let

$$\mathbb{T} = \{\langle a_1, a_1, a_1 \rangle, \ldots, \langle a_n, a_n \, a_n \rangle\}$$

be the set of all n^3 triples over the tokens in A. Given a search database $\mathfrak{r}$ in TNF with tuples $\langle t_1, r_1, a_1, v_1 \rangle, \ldots, \langle t_\ell, r_m, a_m, v_m \rangle$, define $\bar{\mathfrak{r}}$ to be the n^3-vector $\langle \mathfrak{r}_1, \ldots, \mathfrak{r}_{n^3} \rangle$ where $\mathfrak{r}_i$ equals the number of occurrences of the ith triple of $\mathbb{T}$ in the list

$$\langle r_1, a_1, v_1 \rangle, \ldots, \langle r_m, a_m, v_m \rangle.$$

This term vector view on databases leads to several natural search heuristics. The standard Euclidean distance in term vector space from state $\mathfrak{n}$ to target state $\mathfrak{t}$ gives us a *Euclidean heuristic* measure:

$$h_E(\mathfrak{n}) = round\left(\sqrt{\sum_{i=1}^{n^3} (\mathfrak{n}_i - \mathfrak{t}_i)^2} \right)$$

where $\mathfrak{r}_i$ is the ith element of the database vector $\bar{\mathfrak{r}}$.

Normalizing the vectors for state $\mathfrak{n}$ and target $\mathfrak{t}$ gives a *normalized Euclidean heuristic* for the distance between $\mathfrak{n}$ and $\mathfrak{t}$:

$$h_{|E|}(\mathfrak{n}) = round\left(k \sqrt{\sum_{i=1}^{n^3} \left[\frac{\mathfrak{n}_i}{|\bar{\mathfrak{n}}|} - \frac{\mathfrak{t}_i}{|\bar{\mathfrak{t}}|} \right]^2} \right)$$

where $k \geqslant 1$ is a scaling constant and $|\bar{\mathfrak{r}}| = \sqrt{\sum_{i=1}^{n^3} \mathfrak{r}_i^2}$, as usual.

Databases as Term Vectors: Cosine Similarity. Viewing databases as vectors, we can also define a *cosine similarity heuristic* measure, with scaling constant $k \geqslant 1$:

$$h_{\cos}(\mathfrak{n}) = round\left(k\left[1 - \frac{\sum_{i=1}^{n^3} \mathfrak{n}_i t_i}{|\bar{\mathfrak{n}}||\bar{t}|}\right]\right)$$

Cosine similarity measures the cosine of the angle between two vectors in the database vector space. If $\mathfrak{n}$ is very similar to the target t in this space, then $h_{\cos}$ returns a low estimate of the distance between them.

3.5 Supporting Complex Semantic Mappings

The mapping operators in the language $\mathscr{L}$ (Table 1) accommodate dynamic data-metadata structural transformations in addition to simple one-to-one schema matchings. However, as discussed in Section 2.3, many mappings involve complex semantic transformations [14,34,62,64]. As examples of such mappings, consider several basic complex mappings for bridging semantic differences between two tables.

Example 12. A semantic mapping f_1 from airline names to airline ID numbers:

Carrier		CID
AirEast	$\overset{f_1}{\longmapsto}$	123
JetWest		456

A complex function f_2 which returns the concatenation of passenger first and last names:

Last	First		Passenger
Smith	John	$\overset{f_2}{\longmapsto}$	John Smith
Doe	Jane		Jane Doe

The complex function f_3 between FlightsB and FlightsC which maps AgentFee and Cost to TotalCost:

CID	Route	Cost	AgentFee		CID	Route	TotalCost
123	ATL29	100	15		123	ATL29	115
456	ATL29	200	16	$\overset{f_3}{\longmapsto}$	456	ATL29	216
123	ORD17	110	15		123	ORD17	125
456	ORD17	220	16		456	ORD17	236

□

Other examples include functions such as date format, weight, and international financial conversions, and semantic functions such as the mapping from employee name to social security number (which can not be generalized from examples), and so on.

Support for Semantic Mapping Expressions. Any complex semantic function is unique to a particular information sharing scenario. Incorporating such functions in a non-ad hoc manner is essential for any general data mapping solution. Although there has been research on discovering specific complex semantic functions [12,34], no general approach has been proposed which accommodates these functions in larger mapping expressions.

Tupelo supports discovery of mapping expressions with such complex semantic mappings in a straight-forward manner without introducing any specialized domain knowledge. We can cleanly accommodate these mappings in the system by extending $\mathscr{L}$ with a new operator λ which is parameterized by a complex function f and its input-output signature:

$$\lambda^{\text{B}}_{f,\bar{\text{A}}}(R).$$

Example 13. As an illustration of the operator, the mapping expression to apply function f_3 in Example 12 to the values in the `Cost` and `AgentFee` attributes, placing the output in attribute `TotalCost`:

$$\lambda^{\texttt{TotalCost}}_{f_3,\texttt{Cost, AgentFee}}(\texttt{FlightsB}).$$

This is precisely the semantic transformation used in Example 7.   □

The semantics of λ is as follows: for each tuple t in relation R, apply the mapping f to the values of t on attributes $\bar{\text{A}} = \langle \text{A}_1, \ldots, \text{A}_n \rangle$ and place the output in attribute B. The operator is well defined for any tuple t of appropriate schema (i.e., appropriate type), and is the identity mapping on t otherwise. Note that this semantics is independent of the actual mechanics of the function f. Function symbols are assumed to come from a countably infinite set $\mathscr{F} = \{f_i\}_{i=0}^{i=\infty}$.

Discovery of Semantic Mapping Expressions. Tupelo generates data mapping expressions in $\mathscr{L}$. Extending $\mathscr{L}$ with the λ operator allows for the discovery of mapping expressions with arbitrary complex semantic mappings. Given critical input/output instances and indications of complex semantic correspondences f between attributes $\bar{\text{A}}$ in the source and attribute B in the target, the search is extended to generate appropriate mapping expressions which also include the λ operator (Figure 8).

For the purpose of searching for mapping expressions, λ expressions are treated just like any of the other operators. During search all that needs to be checked is that the applications of functions are well-typed. The system does not need any special semantic knowledge about the symbols in $\mathscr{F}$; they are treated simply as "black boxes" during search. The actual "meaning" of a function f, maintained perhaps as a stored procedure, is retrieved during the execution of the mapping expression on a particular database instance. Apart from what can be captured in search heuristics, this is probably the best that can be hoped for in general semantic integration. That is, all data semantics from some external sources of domain knowledge must be either encapsulated in the functions f or somehow introduced into the search mechanism, for example via search heuristics.

This highlights a clear separation between semantic functions which interpret the symbols in the database, such as during the application of functions in $\mathscr{F}$, and syntactic, structural transformations, such as those supported by generic languages like $\mathscr{L}$. This separation also extends to a separation of labor in data mapping discovery, as indicated in our system design framework of Section 2.3: discovering particular complex semantic functions and generating executable data mapping expressions are treated as two separate issues in Tupelo.

Discovering complex semantic functions is a difficult research challenge. Some recent efforts have been successful in automating the discovery of restricted classes of complex functions [12,34]. There has also been some initial research on optimization of mapping expressions which contain executable semantic functions [10].

Focusing on the discovery of data mapping expressions, Tupelo assumes that the necessary complex functions between the source and target schemas have been discovered and that these correspondences are articulated on the critical instance inputs to the system. These correspondences can be easily indicated by a user via a visual interface, such as those discussed in Section 3.2. Internally, complex semantic maps are just encoded as strings in the VALUE column of the TNF relation. This string indicates the input/output type of the function, the function name, and the example function values articulated in the input critical instance.

3.6   Empirical Evaluation

The Tupelo system has been fully implemented in Scheme. In this section we discuss extensive experimental evaluations of the system on a variety of synthetic and real world data sets. Our aim in these experiments was to explore the interplay of the IDA and RBFS algorithms with the seven heuristics described in Section 3.4. We found that overall RBFS had better performance than IDA. We also found that heuristics h_1, h_3, normalized Euclidean, and Cosine Similarity were the best performers on the test data sets.

Experimental Setup. All evaluations were performed on a Pentium 4 (2.8 GHz) with 1.0 GB main memory running Gentoo Linux (kernel 2.6.11-gentoo-r9) and Chez Scheme (v6.9c). In all experiments, the performance measure is the number of states examined during search. We also included the performance of heuristic h_0 for comparison with the other heuristics. This heuristic is constant on all values ($\forall \mathfrak{n}, h_0(\mathfrak{n}) = 0$) and hence induces brute-force blind search (comparable to breadth-first search for both IDA and RBFS [45]). Through extensive empirical evaluation of the heuristics and search algorithms on the data sets described below, we found that the following values for the heuristic scaling constants k give overall optimal performance:

	Norm. Euclidean	Cosine Sim.	Levenshtein
IDA	$k = 7$	$k = 5$	$k = 11$
RBFS	$k = 20$	$k = 24$	$k = 15$

These constant k values were used in all experiments presented below.

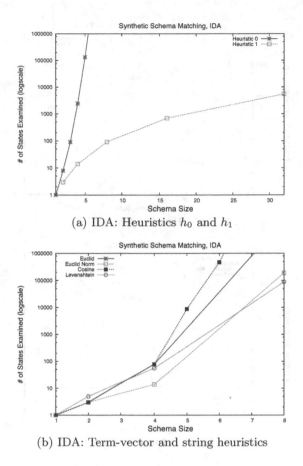

(a) IDA: Heuristics h_0 and h_1

(b) IDA: Term-vector and string heuristics

Fig. 9. Experiment 1: Number of states examined using IDA for schema matching on synthetic schemas

Experiment 1: Mapping on Synthetic Data. In the first experiment, we measured the performance of IDA and RBFS using all seven heuristics on a simple schema matching task.

Data Set. Pairs of schemas with $n = 2, \ldots, 32$ attributes were synthetically generated and populated with one tuple each illustrating correspondences between each schema:

$$\left\langle \frac{A1}{a1}, \frac{B1}{a1} \right\rangle \left\langle \frac{A1\ A2}{a1\ a2}, \frac{B1\ B2}{a1\ a2} \right\rangle \cdots \left\langle \frac{A1\ \cdots\ A32}{a1\ \cdots\ a32}, \frac{B1\ \cdots\ B32}{a1\ \cdots\ a32} \right\rangle$$

Each algorithm/heuristic combination was evaluated on generating the correct matchings between the schemas in each pair (i.e., A1↔B1, A2↔B2, etc.).

Results. The performance of IDA on this data set is presented in Figure 9, and the performance of RBFS is presented in Figure 10. Heuristic h_2 performed

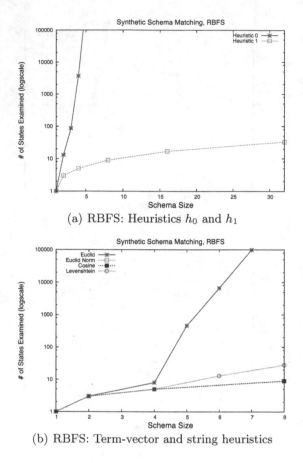

(a) RBFS: Heuristics h_0 and h_1

(b) RBFS: Term-vector and string heuristics

Fig. 10. Experiment 1: Number of states examined using RBFS for schema matching on synthetic schemas

identically to h_0, and heuristic h_3's performance was identical to h_1. Hence they are omitted in Figures 9(a) & 10(a). RBFS had performance superior to IDA on these schemas, with the h_1, Levenshtein, normalized Euclidean, and Cosine Similarity heuristics having best performance.

Experiment 2: Mapping on the Deep Web. In the second experiment we measured the performance of IDA and RBFS using all seven heuristics on a set of over 200 real-world query schemas extracted from deep web data sources.

Data Set. The Books, Automobiles, Music, and Movies (BAMM) data set from the UIUC Web Integration Repository[13] contains 55, 55, 49, and 52 schemas from deep web query interfaces in the Books, Automobiles, Music, and Movies domains, respectively. The schemas each have between 1 and 8 attributes. In this

---

[13] http://metaquerier.cs.uiuc.edu/repository

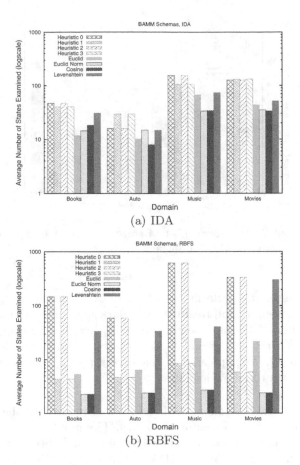

Fig. 11. Experiment 2: Average number of states examined for mapping discovery in the four BAMM domains

experiment, we populated the schemas of each domain with critical instances. We then measured the average cost of mapping from a fixed schema in each domain to each of the other schemas in that domain.

Results. The average performance of IDA on each of the BAMM domains is presented in Figure 11(a). Average RBFS performance on each of the BAMM domains is given in Figure 11(b). The average performance of both algorithms across all BAMM domains is given in Figure 12. We found that RBFS typically examined fewer states on these domains than did IDA. Overall, we also found that the Cosine Similarity and normalized Euclidean heuristics had the best performance.

Experiment 3: Complex Semantic Mapping. In the third experiment we evaluated the performance of Tupelo on discovering complex semantic mapping expressions for real-world data sets in the real estate and business inventory domains.

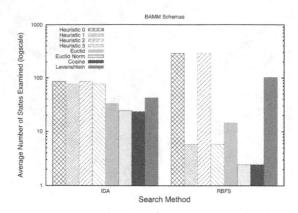

Fig. 12. Experiment 2: Average number of states examined by IDA and RBFS for mapping discovery across all BAMM domains

Data Set. We measured performance of complex semantic mapping with the schemas for the Inventory and Real Estate II data sets from the Illinois Semantic Integration Archive.[14] In the Inventory domain there are 10 complex semantic mappings between the source and target schemas, and in the Real Estate II domain there are 12. We populated each source-target schema pair with critical instances built from the provided datasets.

Results. The performance on both domains was essentially the same, so we present the results for the Inventory schemas. The number of states examined for mapping discovery in this domain for increasing numbers of complex semantic functions is given in Figure 13(a) for IDA and in Figure 13(b) for RBFS. On this data, we found that RBFS and IDA had similar performance. For the heuristics, the best performance was obtained by the h_1, h_3 and cosine similarity heuristics.

Discussion of Empirical Results. The goal of the experiments discussed in this section was to measure the performance of Tupelo on a wide variety of schemas. We found that Tupelo was effective for discovering mapping expressions in each of these domains, even with the simple heuristic search algorithms IDA and RBFS. It is clear from these experiments that RBFS is in general a more effective search algorithm than IDA. Although we found that heuristic h_1 exhibited consistently good performance, it is also clear that there was no perfect all-purpose search heuristic. Tupelo has also been validated and shown effective for examples involving the data-metadata restructurings illustrated in Figure 1 [19]. It was found in that domain that no particular heuristic had consistently superior performance. We can conclude from these observations that work still needs to be done on developing more sophisticated search heuristics.

---

[14] http://anhai.cs.uiuc.edu/archive/

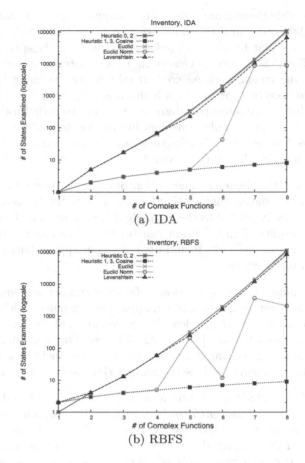

(a) IDA

(b) RBFS

Fig. 13. Experiment 3: Number of states for complex semantic mapping discovery in the Inventory domain

3.7 Related Work

The problem of overcoming structural and semantic heterogeneity has a long history in the database [14] and AI [62] research communities. In Section 2 we have already situated Tupelo in the general research landscape of the data mapping problem. We now briefly highlight related research not discussed elsewhere in the paper.

Schema Matching. A wide variety of existing systems have leveraged AI and machine learning techniques for solving different aspects of schema matching and mapping. These include neural networks, Bayesian learning, and genetic programming approaches [13,51,60,66]. The Tupelo view on data mapping as search complements this body of research. As we have discussed, $\mathscr{L}$ includes (direct) schema matching as a special case and hence Tupelo introduces a novel approach to this research problem.

Data-Metadata Transformations. Few data mapping systems have considered the data-metadata structural transformations used in the Tupelo mapping language $\mathscr{L}$. Systems that have considered some aspects of these transformations include [7,12,59,81]. The systems closest to Tupelo in terms of metadata transformations are the recent work on contextual schema matching by Bohannon et al. [7] and composite approach to schema mapping by Xu and Embley [81], wherein the schema matching paradigm is extended to structural transformations. This work, however, only considers limited subsets of the $\mathscr{L}$ structural transformations. Furthermore, Tupelo complements these systems with the novel perspective of mapping discovery as search.

Example-Driven Data Mapping. The notion of example-based data mapping is an ancient idea, by some accounts dating back at least to the 4th century [63]. Recent work most closely related to the example driven approach of Tupelo include [50,63,66]. The results of this Section can be viewed as contributions towards extending this line of research, including a richer mapping language and new perspectives on mapping discovery.

Executable Mapping Expressions. Most schema matching systems do not address the issue of generating executable mapping expressions, which is in general considered to be an open hard problem [56]. Several notable systems that do generate such expressions include [4,25,31,32,57,59,66]. Our contributions to this research area include extending mapping expressions to include semantic transformations and introducing the Rosetta Stone Principle for expression discovery.

The Data Exchange Problem. A problem closely related to the data mapping problem is the *data exchange problem* [44], proposed to formalize aspects of the Clio schema mapping system developed at IBM [31,59]. This framework has been recently extended to consider the case of data-metadata transformations [35]. Briefly, the data exchange problem is as follows: given a source schema S, target schema T, source instance I, and a set $\Sigma_{S,T}$ of source-to-target dependencies in some logical formalism, find a target instance J that satisfies $\Sigma_{S,T}$ [44]. Fagin et al. have characterized solutions to the data exchange problem and have explored query answering in data exchange settings [44]. A limitation of these results is a restriction of the logical formalism for expressing $\Sigma_{S,T}$ to fragments of first order logic which do not always adequately express naturally occurring data mappings. Furthermore, in data exchange it is assumed (1) that these dependencies are given as input and (2) the target schema T is fixed. In the data *mapping* problem we are concerned precisely with discovering meaningful source to target constraints, given S, T, and perhaps (I, J) as input where the target schema T is potentially dynamic, as we saw in the mapping from FlightsB to FlightsA (Figure 1), which creates as many new route attributes in FlightsA as there are Route values in FlightsB.

In summary, Tupelo complements and extends the research in each of these areas by (1) attacking the data mapping problem as a basic search problem in a state space and by (2) addressing a broader class of mapping expressions including data-metadata transformations and complex semantics functions. We reiterate that, to the best of our knowledge, Tupelo is the first system for data mapping to take an end-to-end modular perspective on the problem and generate such a broad class of database transformations.

Remarks. In this Section we presented and illustrated the effectiveness of the Tupelo system for discovering data mapping expressions between relational data sources. Novel aspects of the system include (1) example-driven generation of mapping expressions which include data-metadata structural transformations and complex semantic mappings and (2) viewing the data mapping problem as fundamentally a search problem in a well defined search space. Mapping discovery is performed in Tupelo using only the syntax and structure of the input examples without recourse to any domain-specific semantic knowledge. The implementation of Tupelo was described and the viability of the approach illustrated on a variety of synthetic and real world schemas. We concluded the Section with an overview of related research results.

4   Concluding Remarks

In this paper, we have studied the long-standing problem of mapping discovery for data sharing and coordination between autonomous data sources. Our aim was the development of a robust perspective on the generic design space of data mapping solutions. In our investigations, we strove towards a deeper understanding of both what data mappings are and how to effectively go about discovering them. The contributions of this paper advance the state of the art of both theoretical and engineering aspects of the data mapping problem, and provide foundations for further progress on both of these fronts. We close by indicating some of the research directions supported by these foundations.

Abstract Formalisms for Data Mapping. We have taken a "set-theoretic" perspective in our development of DMP which lends itself to direct specification depending on problem context (e.g., our definition of RelationalDMP for relational sources in Section 2.3). There are several other high-level frameworks, however, which take a "categorical" perspective on data objects and their mappings. Examples of such abstract frameworks include information-flow theory [38] and institutions [27]. Although these formalisms were not proposed with an eye towards mapping discovery for structured data sources, it would be quite interesting to investigate the connections and disconnections between these perspectives and that of Section 2.

The Tupelo Data Mapping System. There are several promising avenues for future work on Tupelo (Section 3). As is evident from the empirical evaluation presented in Section 3.6, further research remains on developing more sophisticated search heuristics. The Levenshtein, Euclidean, and Cosine Similarity based search heuristics mostly focus on the content of database states. Successful heuristics must measure both content and structure. Is there a good multi-purpose search heuristic? Recently, Gillis and Van den Bussche have studied search heuristics for discovering queries involving negation [25]. Also, we have only applied straightforward approaches to search with the IDA and RBFS algorithms. Further investigation of search techniques developed in the AI literature is warranted. Finally, the perspective of data mapping as search is not limited to relational data sources. In particular, the architecture of the Tupelo system can be applied to the generation of mapping expressions in other mapping languages and for other data models. Based on the viability of the system for relational data sources, this is a promising avenue for future research.

Acknowledgments. We thank the reviewers, the members of the Indiana University database group, and Jan Van den Bussche for their support and many helpful comments.

References

1. Abiteboul, S., Hull, R., Vianu, V.: Foundations of Databases. Addison-Wesley, Reading (1995)
2. Agrawal, R., Somani, A., Xu, Y.: Storage and Querying of E-Commerce Data. In: VLDB, Rome, Italy, pp. 149–158 (2001)
3. Batini, C., Lenzerini, M., Navathe, S.B.: A Comparative Analysis of Methodologies for Database Schema Integration. ACM Comput. Surv. 18(4), 323–364 (1986)
4. Bernstein, P.A., Melnik, S., Mork, P.: Interactive Schema Translation with Instance-Level Mappings. In: VLDB, Trondheim, Norway, pp. 1283–1286 (2005)
5. Berry, M.W., Drmač, Z., Jessup, E.R.: Matrices, Vector Spaces, and Information Retrieval. SIAM Review 41(2), 335–362 (1999)
6. Bilke, A., Naumann, F.: Schema Matching using Duplicates. In: IEEE ICDE, Tokyo, Japan, pp. 69–80 (2005)
7. Bohannon, P., Elnahrawy, E., Fan, W., Flaster, M.: Putting Context into Schema Matching. In: VLDB, Seoul, Korea, pp. 307–318 (2006)
8. Bossung, S., Stoeckle, H., Grundy, J.C., Amor, R., Hosking, J.G.: Automated Data Mapping Specification via Schema Heuristics and User Interaction. In: IEEE ASE, Linz, Austria, pp. 208–217 (2004)
9. Calvanese, D., Giacomo, G.D., Lenzerini, M., Rosati, R.: Logical Foundations of Peer-To-Peer Data Integration. In: ACM PODS, Paris, France, pp. 241–251 (2004)
10. Carreira, P., Galhardas, H.: Execution of Data Mappers. In: ACM SIGMOD Workshop IQIS, Paris, France, pp. 2–9 (2004)
11. Dalvi, N.N., Suciu, D.: Management of Probabilistic Data: Foundations and Challenges. In: PODS, Beijing, pp. 1–12 (2007)
12. Dhamankar, R., Lee, Y., Doan, A., Halevy, A.Y., Domingos, P.: iMAP: Discovering Complex Mappings between Database Schemas. In: ACM SIGMOD, Paris, France, pp. 383–394 (2004)
13. Doan, A., Domingos, P., Halevy, A.: Learning to Match the Schemas of Databases: A Multistrategy Approach. Machine Learning 50(3), 279–301 (2003)
14. Doan, A., Noy, N.F., Halevy, A.Y.: Special Issue on Semantic Integration. SIGMOD Record 33(4) (2004)
15. Eco, U.: The Search for the Perfect Language. Blackwell, Oxford (1995)
16. Euzenat, J., et al.: State of the Art on Ontology Alignment. Technical Report D2.2.3, IST Knowledge Web NoE (2004)
17. Feng, Y., Goldstone, R.L., Menkov, V.: A Graph Matching Algorithm and its Application to Conceptual System Translation. Int. J. AI Tools 14(1-2), 77–100 (2005)

18. Fletcher, G.H.L., Gyssens, M., Paredaens, J., Van Gucht, D.: On the Expressive Power of the Relational Algebra on Finite Sets of Relation Pairs. IEEE Trans. Knowl. Data Eng. 21(6), 939–942 (2009)
19. Fletcher, G.H.L., Wyss, C.M.: Mapping Between Data Sources on the Web. In: IEEE WIRI, Tokyo, Japan, pp. 173–178 (2005)
20. Fletcher, G.H.L., Wyss, C.M.: Data Mapping as Search. In: Ioannidis, Y., Scholl, M.H., Schmidt, J.W., Matthes, F., Hatzopoulos, M., Böhm, K., Kemper, A., Grust, T., Böhm, C. (eds.) EDBT 2006. LNCS, vol. 3896, pp. 95–111. Springer, Heidelberg (2006)
21. Fletcher, G.H.L., Wyss, C.M., Robertson, E.L., Van Gucht, D.: A Calculus for Data Mapping. ENTCS 150(2), 37–54 (2006)
22. Gal, A.: On the Cardinality of Schema Matching. In: Meersman, R., Tari, Z., Herrero, P. (eds.) OTM-WS 2005. LNCS, vol. 3762, pp. 947–956. Springer, Heidelberg (2005)
23. Gal, A.: Why is Schema Matching Tough and What Can We Do About It?. SIGMOD Record 35(4), 2–5 (2006)
24. Garcia-Molina, H.: Web Information Management: Past, Present, Future. In: ACM WSDM, Palo Alto, CA (2008)
25. Gillis, J., Van den Bussche, J.: Induction of relational algebra expressions. In: ILP, Leuven (2009)
26. Giunchiglia, F., Shvaiko, P.: Semantic Matching. Knowledge Eng. Review 18(3), 265–280 (2003)
27. Goguen, J.A.: Information Integration in Institutions. In: Moss, L. (ed.) Jon Barwise Memorial Volume. Indiana University Press (2006)
28. Gottlob, G., Koch, C., Baumgartner, R., Herzog, M., Flesca, S.: The Lixto Data Extraction Project - Back and Forth between Theory and Practice. In: ACM PODS, Paris, France, pp. 1–12 (2004)
29. Grahne, G., Kiricenko, V.: Towards an Algebraic Theory of Information Integration. Information and Computation 194(2), 79–100 (2004)
30. Grundy, J.C., Hosking, J.G., Amor, R., Mugridge, W.B., Li, Y.: Domain-Specific Visual Languages for Specifying and Generating Data Mapping Systems. J. Vis. Lang. Comput. 15(3-4), 243–263 (2004)
31. Haas, L.M., Hernández, M.A., Ho, H., Popa, L., Roth, M.: Clio Grows Up: From Research Prototype to Industrial Tool. In: ACM SIGMOD, Baltimore, MD, pp. 805–810 (2005)
32. Habegger, B.: Mapping a Database into an Ontology: a Relational Learning Approach. In: IEEE ICDE, Istanbul, pp. 1443–1447 (2007)
33. Harris, R.: The Language Connection: Philosophy and Linguistics. Thoemmes Press, Bristol (1997)
34. He, B., Chang, K.C.-C., Han, J.: Discovering Complex Matchings Across Web Query Interfaces: a Correlation Mining Approach. In: ACM KDD, Seattle, WA, pp. 148–157 (2004)
35. Hernández, M.A., Papotti, P., Tan, W.-C.: Data Exchange with Data-Metadata Translations. In: VLDB, Auckland, New Zealand (2008)
36. Hull, R.: Managing Semantic Heterogeneity in Databases: a Theoretical Perspective. In: ACM PODS, Tucson, AZ, pp. 51–61 (1997)
37. Jain, M.K., Mendhekar, A., Van Gucht, D.: A Uniform Data Model for Relational Data and Meta-Data Query Processing. In: COMAD, Pune, India (1995)
38. Kalfoglou, Y., Schorlemmer, M.: Ontology Mapping: the State of the Art. Knowledge Eng. Review 18(1), 1–31 (2003)

39. Kashyap, V., Sheth, A.: Semantic and Schematic Similarities Between Database Objects: A Context-Based Approach. VLDB J. 5(4), 276–304 (1996)
40. Kedad, Z., Xue, X.: Mapping Discovery for XML Data Integration. In: Meersman, R., Tari, Z. (eds.) OTM 2005. LNCS, vol. 3760, pp. 166–182. Springer, Heidelberg (2005)
41. Kementsietsidis, A., Arenas, M., Miller, R.J.: Mapping Data in Peer-to-Peer Systems: Semantics and Algorithmic Issues. In: ACM SIGMOD, San Diego, CA, pp. 325–336 (2003)
42. Kent, W.: The Unsolvable Identity Problem. In: Extreme Markup Languages, Montréal, Quebec, Canada (2003)
43. Kim, W., Seo, J.: Classifying Schematic and Data Heterogeneity in Multidatabase Systems. IEEE Computer 24(12), 12–18 (1991)
44. Kolaitis, P.G.: Schema Mappings, Data Exchange, and Metadata Management. In: ACM PODS, Baltimore, MD, pp. 61–75 (2005)
45. Korf, R.E.: Depth-First Iterative-Deepening: An Optimal Admissible Tree Search. Artif. Intell. 27(1), 97–109 (1985)
46. Korf, R.E.: Linear-Space Best-First Search. Artif. Intell. 62(1), 41–78 (1993)
47. Krishnamurthy, R., Litwin, W., Kent, W.: Language Features for Interoperability of Databases with Schematic Discrepancies. In: ACM SIGMOD, Denver, CO, pp. 40–49 (1991)
48. Lenzerini, M.: Data Integration: A Theoretical Perspective. In: ACM PODS, Madison, WI, pp. 233–246 (2002)
49. Levenshtein, V.I.: Dvoichnye Kody s Ispravleniem Vypadenii, Vstavok i Zameshchenii Simvolov. Doklady Akademii Nauk SSSR 163(4), 845–848 (1965)
50. Levy, A.Y., Ordille, J.J.: An Experiment in Integrating Internet Information Sources. In: AAAI Fall Symposium on AI Applications in Knowledge Navigation and Retrieval, Cambridge, MA, pp. 92–96 (1995)
51. Li, W.-S., Clifton, C.: SEMINT: A Tool for Identifying Attribute Correspondences in Heterogeneous Databases Using Neural Networks. Data & Knowl. Eng. 33(1), 49–84 (2000)
52. Litwin, W.: Bridging a Great Divide: Past, Present, and Future in Multidatabase Interoperability. In: InterDB, Namur, Belgium (2005)
53. Litwin, W., Ketabchi, M.A., Krishnamurthy, R.: First Order Normal Form for Relational Databases and Multidatabases. SIGMOD Record 20(4), 74–76 (1991)
54. Litwin, W., Mark, L., Roussopoulos, N.: Interoperability of Multiple Autonomous Databases. ACM Comput. Surv. 22(3), 267–293 (1990)
55. Matuszek, C., Cabral, J., Witbrockand, M., DeOliveira, J.: An Introduction to the Syntax and Content of Cyc. In: Baral, C. (ed.) Technical Report SS-06-05, pp. 44–49. AAAI Press, Menlo Park (2006)
56. Melnik, S.: Generic Model Management: Concepts and Algorithms. Springer, Berlin (2004)
57. Melnik, S., Bernstein, P.A., Halevy, A.Y., Rahm, E.: Supporting Executable Mappings in Model Management. In: ACM SIGMOD, Baltimore, MD, pp. 167–178 (2005)
58. Miller, R.J.: Using Schematically Heterogeneous Structures. In: ACM SIGMOD, Seattle, WA, pp. 189–200 (1998)
59. Miller, R.J., Haas, L.M., Hernández, M.A.: Schema Mapping as Query Discovery. In: VLDB, Cairo, Egypt, pp. 77–88 (2000)
60. Morishima, A., Kitagawa, H., Matsumoto, A.: A Machine Learning Approach to Rapid Development of XML Mapping Queries. In: IEEE ICDE, Boston, MA, pp. 276–287 (2004)

61. Nilsson, N.J.: Artificial Intelligence: A New Synthesis. Morgan Kaufmann, San Francisco (1998)
62. Noy, N.F., Doan, A., Halevy, A.Y.: Special Issue on Semantic Integration. AI Magazine 26(1) (2005)
63. Perkowitz, M., Doorenbos, R.B., Etzioni, O., Weld, D.S.: Learning to Understand Information on the Internet: An Example-Based Approach. J. Intell. Inf. Syst. 8(2), 133–153 (1997)
64. Rahm, E., Bernstein, P.A.: A Survey of Approaches to Automatic Schema Matching. VLDB J. 10(4), 334–350 (2001)
65. Raman, V., Hellerstein, J.M.: Potter's Wheel: An Interactive Data Cleaning System. In: VLDB, Roma, Italy, pp. 381–390 (2001)
66. Schmid, U., Waltermann, J.: Automatic Synthesis of XSL-Transformations from Example Documents. In: IASTED AIA, Innsbruck, Austria, pp. 252–257 (2004)
67. Sheth, A.P., Larson, J.A.: Federated Database Systems for Managing Distributed, Heterogeneous, and Autonomous Databases. ACM Comput. Surv. 22(3), 183–236 (1990)
68. Shu, N.C., Housel, B.C., Taylor, R.W., Ghosh, S.P., Lum, V.Y.: EXPRESS: a Data EXtraction, Processing, and Restructuring System. ACM Trans. Database Syst. 2(2), 134–174 (1977)
69. Shvaiko, P., Euzenat, J.: A Survey of Schema-Based Matching Approaches. In: Spaccapietra, S. (ed.) Journal on Data Semantics IV. LNCS, vol. 3730, pp. 146–171. Springer, Heidelberg (2005)
70. Stuckenschmidt, H., van Harmelen, F.: Information Sharing on the Semantic Web. Springer, Berlin (2005)
71. Wache, H., Vögele, T., Visser, U., Stuckenschmidt, H., Schuster, G., Neumann, H., Hübner, S.: Ontology-based integration of information – a survey of existing approaches. In: IJCAI (2001)
72. Wang, G., Goguen, J.A., Nam, Y.-K., Lin, K.: Critical Points for Interactive Schema Matching. In: Yu, J.X., Lin, X., Lu, H., Zhang, Y. (eds.) APWeb 2004. LNCS, vol. 3007, pp. 654–664. Springer, Heidelberg (2004)
73. Warren, R.H., Tompa, F.W.: Multi-Column Substring Matching for Database Schema Translation. In: VLDB, Seoul, Korea, pp. 331–342 (2006)
74. Wiederhold, G.: The Impossibility of Global Consistency. OMICS 7(1), 17–20 (2003)
75. Wiesman, F., Roos, N.: Domain Independent Learning of Ontology Mappings. In: AAMAS, New York, NY, pp. 846–853 (2004)
76. Winkler, W.E.: The State of Record Linkage and Current Research Problems. Technical Report RR99/04, U.S. Bureau of the Census, Statistical Research Division (1999)
77. Wyss, C.M., Robertson, E.L.: A Formal Characterization of PIVOT/UNPIVOT. In: ACM CIKM, Bremen, Germany, pp. 602–608 (2005)
78. Wyss, C.M., Robertson, E.L.: Relational Languages for Metadata Integration. ACM Trans. Database Syst. 30(2), 624–660 (2005)
79. Wyss, C.M., Van Gucht, D.: A Relational Algebra for Data/Metadata Integration in a Federated Database System. In: ACM CIKM, Atlanta, GA, USA, pp. 65–72 (2001)
80. Wyss, C.M., Wyss, F.I.: Extending Relational Query Optimization to Dynamic Schemas for Information Integration in Multidatabases. In: ACM SIGMOD, Beijing (2007)
81. Xu, L., Embley, D.W.: A Composite Approach to Automating Direct and Indirect Schema Mappings. Information Systems 31(8), 697–732 (2006)

Using Semantic Networks and Context in Search for Relevant Software Engineering Artifacts

George Karabatis[1], Zhiyuan Chen[1], Vandana P. Janeja[1], Tania Lobo[1],
Monish Advani[1], Mikael Lindvall[2], and Raimund L. Feldmann[2]

[1] Department of Information Systems, University of Maryland, Baltimore County (UMBC)
1000 Hilltop Circle, Baltimore, MD 21250, USA
[2] Fraunhofer USA Center for Experimental Software Engineering
4321 Hartwick Rd., College Park, MD 20742, USA

Abstract. The discovery of relevant software artifacts can increase software re-use and reduce the cost of software development and maintenance. Furthermore, change requests, which are a leading cause of project failures, can be better classified and handled when all relevant artifacts are available to the decision makers. However, traditional full-text and similarity search techniques often fail to provide the full set of relevant documents because they do not take into consideration existing relationships between software artifacts. We propose a metadata approach with semantic networks[1] which convey such relationships. Our approach reveals additional relevant artifacts that the user might have not been aware of. We also apply contextual information to filter out results unrelated to the user contexts, thus, improving the precision of the search results. Experimental results show that the combination of semantic networks and context significantly improve the precision and recall of the search results.

Keywords: software engineering, search for artifacts, semantic networks, context.

1 Introduction

In the domain of software engineering software changes are inevitable, for example, due to requirements change, but cause several well-known problems if not handled properly. They can lead to severe time pressure and project delays due to underestimation of the scope of the change. Major studies of today's software intensive systems consistently find surprisingly large numbers of failed, late, or excessively expensive systems [36] and according to [29] requirement change is one of the most common causes of software project failure.

Thus, searching for relevant software development artifacts (requirements documents, design documents, source code, etc.), has become increasingly important. For

---

[1] Semantic networks are graphs which represent knowledge by interconnecting nodes through edges. They have been used to describe and classify concepts for many centuries. According to Sowa the earliest known semantic network was drawn in the third century AD by the Greek philosopher Porphyry (Porfyrios) to graphically illustrate the categories of Aristotle (Sowa 1992). For a detailed background on semantic networks see (Sowa).

S. Spaccapietra, L. Delcambre (Eds.): Journal on Data Semantics XIV, LNCS 5880, pp. 74–104, 2009.

example, software developers often need to find out whether there are some similar software components or software designs to better respond to a software change request. Finding such related information may greatly reduce the cost of software change or allow more accurate estimation of the cost of the change, which may lead to better decisions (e.g., whether to accept or reject the change request). One study found that the amount of necessary software changes was three times higher than originally predicted [45], indicating that a better search technology may be one possible solution to this problem. We also believe that our technique can also help in the development of new software.

Importance of Capturing Ad hoc Relationships: There are several obstacles to finding relevant artifacts, especially in the domain of software engineering. First, the relationships between artifacts are often ignored by existing full-text or similarity search technologies [9, 24, 1], but are extremely important for finding relevant software artifacts. For example, many software projects have various versions (thus, these versions are related), a large number of requirements documents (related to the code that implements them), and often implement overlapped functionality (thus, different projects are related to each other). However, it is extremely difficult for someone to find relevant artifacts if that person is unfamiliar with the software project structure and its history including its relationships to other software products and their evolution.

Motivating Example: The following example describes some of these difficulties focusing on the ones that arise during a search for relevant information that would be triggered for example, by a change request. Fig. 1 shows the relationships between two closely related software projects: The Tactical Separation Assisted Flight Environment (TSAFE) and the Flight Management System (FMS). The arrows are directional and indicate that the source node is related to the target node and the number identifies the degree of relevance between the two connected nodes. Details of these two projects can be found in Section 5.

Suppose a developer of FMS version 5 receives a change request to add the capability to change geographical area in run time. The developer tries to find artifacts related to the keyword "FMS 5". Certainly, an existing text search tool such as Google Desktop or a search tool that matches artifacts with similar attributes could be used. Unfortunately, related artifacts such as TSAFE II Loss Of Separation (LOS) (which is functionally equivalent to FMS 5), TSAFE II (without the LOS option), TSAFE II Dynamic Map (which implements the sought functionality in a different project TSAFE), and the requirement document for Dynamic Map (which is the design document for the sought functionality) are unlikely to be retrieved. The reason is that these artifacts do not contain the keyword FMS. In order to overcome this problem, one could make TSAFE a synonym of FMS or define a similarity score between these two terms. However, in this context, FMS has two meanings: Flight Management System and Finance Management System (an accounting system). Thus, a search for TSAFE would retrieve artifacts for both the Flight Management System and Finance Management System. In conclusion, the design information related to TSAFE II DM, which implements the sought functionality, cannot be retrieved without extensive searching due to its distant and indirect relationship to FMS 5 (see Fig. 1).

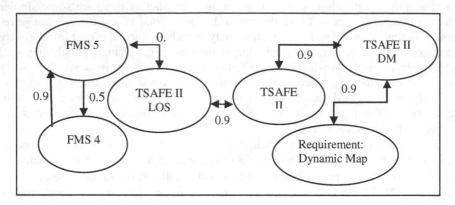

Fig. 1. A (partial) semantic network for Air Traffic Control Software

Our Approach and Contributions: We use two techniques to solve this problem: Semantic networks and context. Next, we give a brief description of these two techniques. We use semantic networks to capture ad hoc relationships between artifacts. Fig. 1 shows a partial view of a semantic network in our example. The nodes represent artifacts and the links represent relationships between artifacts. The number on each edge quantifies the degree of relevance (i.e., the strength of the relationship) of two artifacts. We also infer indirect relevance scores between two indirectly linked artifacts. For example, the relevance score between FMS 5 and TSAFE II DM equals the product of relevance scores between FMS 5 and TSAFE II LOS, TSAFE II LOS and TSAFE II, and TSAFE II and TSAFE II DM.

Now, given a search for information pertinent to FMS 5, we can use the semantic network to add relevant artifacts. A full-text search engine (or a similarity search engine) returns FMS 5 as result. We can then expand the results by adding any artifacts whose relevance score to FMS 5 exceeds a certain user defined threshold (let's say 0.7). This threshold identifies the degree of semantic closeness (relevance score) between related artifacts in the semantic network. For example, only artifacts which are relevant with a relevance score above threshold t are included in the semantic network recommendations. Thus the results will include FMS 5, TSAFE II LOS, TSAFE II, TSAFE II DM, and Requirement: Dynamic Map.

Although semantic networks recommend semantically relevant answers, these answers can fit the user's query more precisely if additional information about the user, the project, the environment, etc. were to be provided. We use context, which contains information relevant to the user (such as user's roles and the current and/or previous projects) to enable the semantic network to target the user queries more accurately. For example, some users may be only interested in requirements or design documents (since they may not be programmers and can not read source code). In our approach, we store different types of context information and use it to filter the results generated by semantic network. In the above example, the context associated with that user, is that the answer must be requirement documents, thus only the requirement for Dynamic Map is returned, which is exactly what the user wants.

We consider each artifact to be a node associated with certain characterizing attributes. We begin by constructing a first mode semantic network, referred to as Similarity based Network, which is a graph that identifies the similarity between the nodes in terms of the features associated with each node. We next utilize semantic rules to augment the first mode network and discover a second mode network, which we call Rule Enforced Network. Although both semantic networks and context have been used individually in search [26] and other applications, there are three important differences in this paper.

First, we use semantic networks to model the ad hoc relationships between artifacts while most existing work such as WordNet [26] and concept maps [74] model relationships between keywords. We find that in many domains such as software engineering, the relationships between artifacts are often difficult to model at the keyword level. For example, it is very difficult to model the relationship between a software release and its subsequent releases at the keyword level. On the other hand, it is straightforward to add such links between artifacts: a simple rule can be created to add such links automatically.

Second, this paper combines semantic network with context. Using a semantic network alone may improve the recall of search results because more artifacts are returned. However, it may not improve precision of the search results because not all these additional artifacts may be considered pertinent by the user. This paper uses context to filter out irrelevant artifacts based on existing contexts. We have conducted experiments which show that the combination of semantic network and context leads to better precision and recall.

Third, there has been little work on searching for software engineering artifacts, which has become increasingly important for the software industry – especially in regard to the increasing number of change requests and extended life cycles of today's software products. To the best of our knowledge, we are not aware of any study that uses both semantic networks and context in the software engineering domain. Specifically our contributions are as follows:

1. We automatically construct a semantic network from artifacts and their associated characterizing attributes.
2. We keep a provision for adding external semantic rules supplied by a domain expert, that when applied to the semantic network, they augment and enhance it.
3. We automatically find, using our semantic network, not only the requested artifacts based on a user query, but additional relevant ones that the user might have not been aware of.
4. We apply context on the result set of the user query to enhance the quality of search over artifacts and include only contextually related artifacts.
5. We demonstrate through experiments, using software artifacts from a software test bed, that the combination of semantic networks and context significantly improve both the precision and recall of search results.

Although this paper focuses on the software engineering domain, we believe the proposed approach is suitable to other domains too. The rest of the paper is organized as follows: Section 2 describes related work and Section 3 gives preliminaries on semantic networks and context. In Section 4 we describe our approach, and in Section 5 we present and discuss validating experiments. Section 6 concludes the paper.

2 Related Work

Related work on software change and reuse. The problem of determining software change impact has a long history. Haney's early model for module connection analysis predicts the propagation of change among modules of a system [33]. It assumes that a change in one module is likely to cause change in other modules. A probability connection matrix subjectively models the dependencies between modules. Our approach is different from Haney's in that we do not only model dependencies between modules, but between all artifacts carrying design information as well as relationships between projects, and we use context.

Many useful theoretical models for impact analysis and change-related processes are collected in the excellent overview by [11]. Different approaches to identify change are described; for instance, traceability analysis for change propagation between work-products, ripple-effect analysis for propagation within work-products, and change history analysis to understand relationships to previous changes. Many approaches address reuse of various artifacts [62, 57, 61]. These approaches make use of metadata (i.e., tags) to describe the artifacts, which are used to classify [57, 63] and retrieve them [56, 7].

Latent Semantic Indexing (LSI), an information retrieval technique, has been used to recover links between various artifacts [47, 50] that share a significant number of words (not necessary the words being searched). However, in the software engineering domain the usefulness of this approach is sometimes limited. As described in our Motivating Example, the critical keywords TSAFE and FMS do not appear in the same artifacts and the description of the change request uses different terminology. Furthermore, the similarities between two artifacts are symmetric in LSI, which is often not true in practice. For example, given a software release, the next and newer release is likely more interesting than the previous one. Since semantic networks do not require the existence of such shared keywords and they do not require that similarities are symmetric, our approach does not have these limitations.

Canfora et al. describe the state of the art of information retrieval and mention three areas in which information retrieval is applied to software maintenance problems: Classifying maintenance request (i.e., change request), finding an expert for a certain maintenance request, and identifying the source that will be impacted by a maintenance request [15]. For example, in [48] the authors use various technologies such as Bayesian classifiers to classify maintenance requests. In [5, 18] [54] the authors determine who is the expert for a certain change request based on who resolved a similar change request in the past based on data from version control systems and try to identify similar change requests from the past. The main difference with our approach is that we model distant relationships connecting projects and artifacts that are similar, but would most likely never show up using similarity-based searchers. The impact on source code from a certain change request has been studied in [14] by correlating change request descriptions with information provided in version management systems such as Bugzilla.

Related work on semantic networks. Semantic Networks have been used in philosophy, psychology and more recently in information and computer science. Sowa gives a descriptive outline on the types and use of semantic networks in different disciplines in [70, 69]. Semantic networks have long been used to represent relationships [51]. Pearl used probabilities in semantic networks and performed extensive work in applying statistics and probability in causal semantic networks [59, 58] to derive such networks from observed data.

There has also been work on discovering semantic similarity in [22] based on generalization/specialization, and positive/negative association between classes; the topic of discovering and ranking semantic relationships for the semantic web is also relevant [3, 67]. Our work is also linked to the specification of relationships among database objects stored in heterogeneous database systems [38, 28, 68, 65]. We have used semantic networks to enhance the results of a user query in different application domains such as the environmental [17] and e-government of water quality management [16]. However these early approaches do not support automated creation of the semantic network and do not incorporate context as part of the solution. We are now applying our approach in the domain of software engineering: see [43] for an early and initial approach describing a limited scope of this problem in software engineering. Quite related is the work on ConceptNET a large scale concept base which captures common sense concepts [46].

Related work on context. A significant part of scientific literature is related to the use of context information related disciplines and in the social sciences such as psychology and sociology. Pomerol and Brezillon examined notions of context and identified its forms as external, contextual, and proceduralized [60]. Bazire and Brezillon made an analysis on 150 definitions of context found on the web, and concluded that the definition of context depends on the field of knowledge it belongs to [8]. For a comprehensive examination of an operational context definition see [78] and for context definitions in artificial intelligence, databases, communication, and machine learning see [13]. Lieberman and Selker presented context in computer systems and described it as "everything that affects the computation except the explicit input and output" [42]. There is related research performed within the scope of data integration and interoperability using context [37, 76, 30, 23]. Context has also been used as an aid in defining semantic similarities between database objects [39].

Sowa provides an overview on facts and context in [71]. Context has been used in multiple settings: Semantic knowledge bases utilize a partial understanding of context; WordNet is such an example, where context is expressed in natural language [26]. It has also been used to provide better algorithms for ranked answers by incorporating context into the query answering mechanism [2] and to improve query retrieval accuracy [66]. Significant work on context-sensitive information retrieval was performed in [66, 72, 35, 27]. However, we focus on how to take context into consideration when using semantic networks. Graphs that represent context have been used to provide focused crawling to identify relevant topical pages on the web [20]. Methods to model and represent context for information fusion from sensors using relational database model are described in [77]. Context has also been used to prune semantic networks to improve performance, by marking and thus using only nodes which are pertinent in specific contexts [31]. In addition, graphs were used by [55] to infer the context and fit it into an existing semantic network. However, in our work we keep the semantic networks separate from context, and we avoid automated inference of context. Finkelstein et al. identify the difficulties in automatic extraction of context, especially with text, as documents may be large, could contain multiple concepts, and could inject a lot of noise [27]. We have decided to collect context either by observing user actions, or explicitly by the user, an approach which is also used in [41] where the context is used by the system; however it differs from our approach as it they use it to perform a rewrite of the original user query, whereas we apply a filtering technique on the expanded result.

Related work on the semantic web and ontologies. A significant amount of work on semantics and the meaning of concepts has been done for the semantic web [10, 75]. The Web Ontology Language OWL [53] has been used to model concepts of a domain that are to be shared by others providing a relevance to the concept of semantic networks. McCarthy introduced the *ist(C,p)* predicate to disambiguate when a proposition *p* is true in context *C* [52] and in [32] the authors adapt the *ist* construct of context and address the contextual problems which arise in aggregation on the semantic web. The restrictions of the standard OWL specification, such that it allows neither directionality of information flow, nor local domain (which is of utmost importance for contexts), nor context mappings, are overcome by extending the syntax and the semantics of OWL via explicit context mappings [12]. The notion of relationships between concepts is also related to the topic maps or concept maps [74]. The major thrust of our work is to create a methodology that utilizes semantic networks and contextual information to support software engineers in their search of relevant artifacts. It can be implemented in a variety of ways:

- As a stand-alone system, as we present in this paper
- On the web, using semantic web technologies, such as OWL and RDF [4]
- In a combination of the above two techniques

The concepts presented in this paper can also be adapted and implemented on the semantic web, for example, expressing relationships using OWL. However, such effort is not within the scope of this paper, but we plan to investigate semantic web technologies in the future.

3 Preliminaries

In this Section we provide an introduction to some topics and notation that are being used in the remainder of the paper, around the concepts of semantic networks and context.

Artifacts. In our software engineering setting, we assume that each artifact is associated with metadata represented as a set of attribute-value pairs. For example, the FMS version 5 has the following attributes: Name = Flight Management System Version 5, Type = Code, Programming Language = Java. In addition to these attributes the artifacts can be parsed to derive additional attributes. For instance in a Java program, import statements, function names etc. also provide valuable information about the artifact and can indeed be used as attributes describing the artifacts.

Semantic networks. A semantic network represents ad hoc relationships among artifacts.

Definition 1 [Semantic Network]. A Semantic Network N(V, E) is a directed graph where V is a set of nodes and E is a set of edges. Each node corresponds to an artifact, and each edge links two relevant artifacts v_i and v_j and has a score $w(v_i, v_j)$ in the range of 0 to 1, representing the degree of relevance between the source artifact and the destination artifact.

Fig. 1 illustrates a partial semantic network for the FMS and the TSAFE projects. The network contains knowledge of multiple people, e.g., an individual programmer of TSAFE may not know the relationships of artifacts in FMS, and vice versa, but a software architect may know that TSAFE is related to FMS, although the software architect may not know in detail the relationships between the artifacts within each system. That is, each of them only has knowledge of a part of the semantic network. However, based on the relevance scores between neighboring nodes in the network, it is possible to infer the relevance between any two nodes (as far apart as FMS 5 is to TSAFE II DM – see Fig. 1). Thus, one can discover more semantically related information compared to individual knowledge. Next we define the relevance score between any two artifacts in the network.

Definition 2 [Relevance Score]. If v_i and v_j are two nodes in a semantic network $N(E, V)$, there are k paths $p_1, ..., p_k$ between v_i and v_j, where path p_l (1 <= l <= k) consists of nodes $v_{l1}, ..., v_{l|p l|+1}$ ($|p_l|$ is the length of path p_l). The relevance score rs as defined by [17] between v_i and v_j is

$$rs = \max\left(\prod_{1 \leq i \leq |pl|} w(v_{l_i}, v_{l_{i+1}}) \right)$$

The above formula computes the relevance score between v_i and v_j as the maximum relevance score of all paths connecting v_i and v_j. The relevance score of such a path is computed using conditional probabilities under the assumption that they are independent.

For instance, the relevance score between 'FMS 5' and 'TSAFE II DM' can be considered as the conditional probability of a software developer interested in the TSAFE II DM given that the developer is interested in the related product line FMS 5. Using the standard notation for conditional probability, we have:

P(TSAFE II DM | FMS 5) = P(TSAFE II DM, TSAFE II, TSAFE II LOS | FMS 5)

because the developer considers that TSAFE II DM and FMS 5 are related if all artifacts on the path from FMS 5 to TSAFE II DM (TSAFE II LOS and TSAFE II) are considered to be related. Using chain rules and assuming all conditional probabilities are independent [64], we have:

P(TSAFE II DM, TSAFE II, TSAFE II LOS | FMS 5) =

P(TSAFE II LOS | FMS 5) × P(TSAFE II | TSAFE II LOS) × P(TSAFE II DM | TSAFE II) = 0.9 × 0.9 × 0.9 = 0.73.

Thus, a developer receiving a change request for FMS 5 and using the semantic network, would be able to find relevant artifacts such as FMS 4, TSAFE II LOS, TSAFE II, LOS Detector requirement, TSAFE II DM, and Requirement: Dynamic Map. Note that we can easily specify "not related" information in the semantic network by simply not adding a link between them. For example, there shall be no link between the Finance Management System and TSAFE. We also assume that relationships between nodes do not have to reflect the same attribute. By design, we just need to have any relationships established between the nodes of the graph, and we do not necessarily need to have probabilities of the same attribute to calculate paths.

Context. We consider context to be significantly important in the search for semantically related information as every single search is performed within a specific context. Although this context may not explicitly appear in the query terms, nevertheless it does exist, and the user expects the system to provide information relevant to this context. In general, users may not be aware of context when they first search for information. However they become cognizant of context when they receive results that are irrelevant or not applicable to the current context, i.e., when a search for FMS returns Finance Management System. A highly beneficial characteristic of our system is that it takes advantage of context in a transparent way to filter and return the most appropriate answers tailored to each user. We consider the following four types of context:

- *User Context* contains information specific to users such as the role of the user (e.g., developers or design analyst), the programming language skills, etc.
- *Application Context* contains information about the application or project the user is working on, such as the name and type of project, etc.
- *Environment Context* includes information about the environment around the user, such as the organization the user belongs to, the operating system of the user's computer, etc.
- *Other Context* is used as a place holder for additional contextual information which does not fit in any of the previous context categories, but still is relevant to the domain of discourse.

Note that there are additional categories of context which do exist, such as security considerations, policies, etc., which are not captured in our system. We acknowledge that it is unlikely to capture all possible types of context and their values in a computer system, since there will always be additional information contributing to context. We limit ourselves to collecting information about the above categories of context, and we do not claim that we can capture all possible context types. For an extensive work on an operational definition of context see [78]. In the domain of software engineering, we claim that such context information is relatively easy to collect as it was the case in our experiments and described below, and assume will be similar in most software engineering settings. Unlike domains such as generic search on the Internet where users submit ad hoc queries and want to find answers immediately, the users in software engineering domain are typically software developers, analysts, project managers, etc., who are regular users of the system and are more willing to provide contextual information in return for more precise search results. User contexts can be gathered by asking the users about past project experience, or by contacting their manager, etc. Application contexts can be obtained by asking the project managers. Environment contexts that are relatively static (i.e., the name of an organization) can be obtained easily, while those that are volatile (i.e., the current software version information) may need additional effort to collect and maintain. However, maintaining contexts falls outside the scope of this paper as we assume that the various types of contexts are already collected, stored, and maintained in a database. Formally a context can be represented in the following format.

Definition 3 [Context]. *The context $C(U_j, T)$ of type T for user U_j, where $T \in$ {User, Application, Environment, Other} is represented as a conjunctive normal form* $\bigwedge_i (L_{i1} \vee L_{i2} \vee \cdots \vee L_{ini})$ *where each L_{ij} is an attribute value pair or its negation* .

For example, the user context of a user who is a design analyst (i.e., interested in design and requirements documents) and does not know C++ is: (Type = Design $\vee$ Type = Requirements) $\wedge \neg$(Programming Language = C++) .

From a systems viewpoint, context is metadata information stored in database tables and it is used in conjunction with semantic networks as follows: Artifacts represented as nodes in semantic networks contain characterizing attributes, which may participate in the attribute value pairs of a context definition. These attributes link semantic networks and context. Relevance related information comes from semantic networks, and in turn is pruned by context-related information through the attribute value pairs. It is important to note that semantic networks and context are somehow orthogonal dimensions, but both use the attributes of the artifacts. Further details are described in Section 4.3.

4 Approach

Our approach consists of the following distinct steps: we first provide a high level overview of the major components of our system and the lifecycle of a user query through the system. Then we present details on the creation of a semantic network: we first derive the universal feature vector which has all the potential attributes across the set of artifacts. Based on the vector a feature vector for each artifact is generated. We utilize the similarities between these feature vectors to generate a similarity based network. This is further enhanced by semantic rules to generate a Rule enhanced network. We also define relevance scores between artifacts. Subsequently we define the context for the semantic network for a more refined result set. Lastly we apply transformation functions on this semantic network. The approach is discussed in the following subsections: Section 4.1 the system overview and the lifecycle of a user query. Section 4.2 describes how we construct semantic networks. Section 4.3 discusses how to use context in our system. Finally, Section 4.4 describes a framework of transformation functions which formalize our overall approach.

4.1 System Overview

In this section we outline the system architecture and the flow of a query from the time it is submitted until the results are returned back to the user.

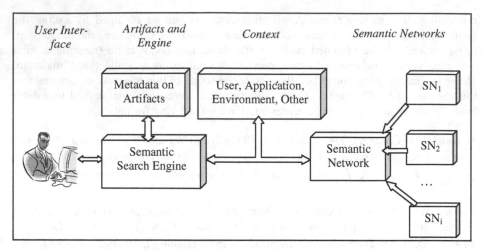

Fig. 2. Conceptual system architecture

A high-level conceptual architecture of our system with its major components is illustrated in Fig. 2. Our prototype system has been implemented using an Oracle database in which we store all types of metadata about software artifacts (attribute-value pairs), semantic networks, and context. Our system stores only metadata, e.g., an identifier (such as a uri) pointing to the location of each actual artifact. A set of semantic networks is depicted to the right part of Fig. 2, each representing a separate software project. All these semantic networks are merged into a larger Semantic Network, which integrates the individual semantic networks into a consolidated one. The edges connecting these networks identify the existence of a potential relationship between them. The strength of this relationship is represented as a relevance score. Another component of our system contains information about the different types of contexts that are collected (User, Application, Environment, etc.) and it is used to identify semantically related information and filters out irrelevant information. Context has been implemented as a set of tables in an Oracle database. Metadata about software artifacts is also stored in the database to be used in the extraction of the initial artifacts based on the user query. The Semantic Search Engine interacts with the users and all major components of our system. It oversees all operations at each component, from the submission of a user query, to its execution, the use of semantic networks and contexts, all the way to displaying the final results to the user.

Semantic networks and context information significantly improve the quality of the query result, since: (1) they enhance the result set with semantically relevant information that the users might not be aware of, and (2) they incorporate contextual knowledge to streamline the result according to user, application and other contexts. We demonstrate the improvement in quality by measuring recall and precision of the results (see Section 5). In this Section we present the lifecycle of a query submitted by a user to our system as illustrated in Fig. 3. Initially, a text search is performed to

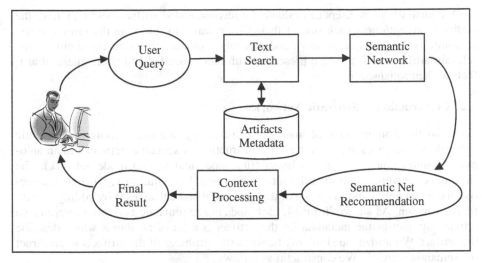

Fig. 3. Life cycle of a user query

collect all related information which matches the user query. This search is performed on a set of metadata about the artifacts[2]. Then, the returned artifacts are given as input to the semantic network. Using the algorithms described in detail in section 4.2 and depending on the value of the user-defined threshold on the relevance score, the semantic network produces an augmented list of recommended artifacts, whose relevance scores to any of the initial artifacts exceed the threshold and thus are semantically relevant to the initial user query. However, this augmented list may not reflect the contextual information pertaining to the user, project, etc. Consequently, the contexts are used next, to filter information accordingly. As a result, only the recommended artifacts which are pertinent to the contexts will be collected and given to the user in the final result set of the original query. For example, assume that a requirement analyst asks a query on "automated collision avoidance," the system first performs a full-text search and returns all artifacts from the database containing these keywords. At this point, the current result set may not contain all relevant artifacts as there could be additional artifacts that are semantically relevant but which are not included in the search results. Then, the system utilizes the semantic network to find all additional artifacts which are semantically related to the current search results; i.e. additional artifacts that do not contain the search keywords explicitly, but are closely related to them (e.g., the Loss of Separation Detection Module, which detects situations where two aircrafts are too close). However, the augmented results containing all semantically relevant artifacts may not be pertinent to the user's context. Therefore, contextual information is extracted from the database and is applied to the set of augmented results to filter out artifacts that are out of context keeping only those that are within context. In our example, only requirement documents (but not source code) are kept in the final result.

---

[2] In this paper, whenever we refer to artifacts we mean the metadata about the artifacts and not the artifacts themselves.

Of course we allow users to evaluate the recommended artifacts and they have the ability to accept/reject each one of them. They can also fine-tune the search query, resubmit to the semantic network and possibly provide a different threshold for the relevant artifacts until they are presented with recommended and contextualized artifacts to their satisfaction.

4.2 Construction of Semantic Networks

To avoid the daunting task of manually constructing and maintaining the semantic network, we adopt an approach for the construction of semantic networks in an automatic manner, consisting of two layers (first mode and second mode network). The first mode network identifies relevant artifacts based on similarity, whereas the second mode network is build on top of the first mode and enhances it by adding semantic information. As shown in Fig. 4, each node in the semantic network represents an artifact and part of the metadata for this artifact is a set of attributes which describe the artifact. We utilize the similarity between the attributes of the artifacts to construct the semantic network. We construct it as follows.

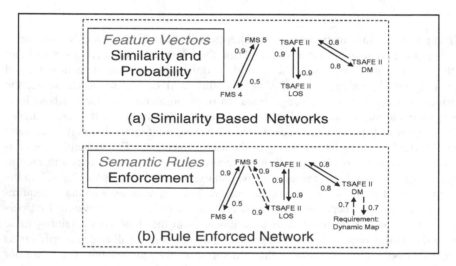

Fig. 4. Creating a semantic network

Automatic Generation of Feature Vectors

Let $X = \{x_1,...x_n\}$ be the set of artifacts, where each $x_i \in X$ is associated with a set of characterizing attributes $a_i = \{a_{i1},....a_{im}\}$. The values of these attributes can be transformed into categorical values (binary) and form a feature vector $f_i = \{f_{i1}, \ldots, f_{im}\}$. In order to automatically create a similarity based network we first need to generate the feature vector associated with each artifact. Our approach is generalizable to continuous attributes such that they can be discretized into categorical variables. Additionally we can also handle textual variables since we can parse the features from code files. These feature vectors are used to determine how similar the artifacts are in terms of the attributes characterizing them and they are utilized to generate the semantic network; therefore, we outline our process and algorithm here for the generation of feature vectors. We discover features by class of artifacts. For instance, software artifacts can be viewed as programs, requirement specifications, test cases and so on. For each such class of artifacts we produce a universal feature vector by creating a parser for the artifacts. This can be seen as a preprocessing step necessary to acquire the data about the artifacts. In the case of Java programs, using a text file parser, we extract features such as import statements and function names, and we add them to the feature vector. The union of all feature vectors creates a universal vector (V) containing all artifacts. Thus $V=\{ a_{11} \cup a_{21},....\cup a_{im} \}$ and it will be used for the similarity based network.

Algorithm 1 The Feature Vector generation algorithm

Require: Set of artifacts X where each $x_i \in$ X
Ensure: Set of artifacts X where each $x_i \in$ X is associated with a set of attributes $a_i = \{a_{i1},..,a_{im}\}$ and a set of features $f_i = \{f_{i1},..,f_{im}\}$
  1: **for** i = 1 to $|X|$ **do**
  2:    {Read artifact x_i}
  3:    {$a_i \leftarrow$ Parse Attributes(x_i)}
  4:    {add(U, a_i)}
  5: **end for**
  6: {Initialize Feature Vector f}
  7: **for** i = 1 to $|X|$ **do**
  8:   **for** z = 1 to $|U|$ **do**
  9:       **if** $u_z \in x_i$ **then**
10:         {$f_{iz} = 1$}
11:      **else**
12:         {$f_{iz} = 0$}
13:      **end if**
14:   **end for**
15: **end for**

Algorithm for Automatic Feature Vector Generation. We outline our algorithm for the generation of the Feature Vector. The algorithm takes as an input the set of artifacts $X=\{x_1,...., x_n\}$. On lines 1-5 we generate a universal vector U, by parsing through the artifacts. This essentially finds all the attributes from the various artifacts, from x_1 to x_n, and stores them in U. So for instance if we are parsing a java program then the import statements will be the attributes of the Universal vector. We then create a feature vector from U on lines 7-15. We parse the artifacts to note the presence or absence of an attribute in the artifact. For example if we have a Java program artifact with an import java.util statement then the feature in the vector for this artifact will have a value 1 vs. another java program without the import statement will have a value 0 for the feature. The parser can be modified to handle other types of languages such as c++, python etc. Programming languages provide a structured environment to handle such a parsing. However documents may not be parsed easily using this method since their structure is not very well defined. The complexity of the algorithm is O(N |U|) where N is the number of artifacts and |U| is the size of the Universal Vector.

Similarity Based Network

Let us assume that we have a set of n artifacts $X = \{x_1,...x_n\}$, where each $x_i \in X$ is associated with a set of m features captured in a feature vector $f_i = \{f_{i1}, ..., f_{im}\}$. We use a Jaccard similarity coefficient[3] to quantify the similarity among the feature vectors of the artifacts. Based on the Jaccard coefficients we connect similar nodes using edges and start creating the semantic network. We add probabilities on the edges as follows: given a pair of nodes x_p and x_q such that there exists a similarity between the two nodes the probability $w(xp, xq)$ of traversing from node x_p to x_q is:

$$w(x_p, x_q) = \frac{J_{pq}}{deg_p} \quad \text{where } deg_p = \sum_{j=1}^{k} J_{pj}$$

J_{pq} is the Jaccard similarity coefficient between the feature vectors of artifacts (nodes) x_p and x_q. J_{pj} is the weighted degree of the node p, and k is the number of incident edges on p. Thus, based on the similarity and probability computations we get a first mode semantic network as shown in Fig. 4(a), which we refer to as *Similarity based Network*. There could be several disconnected first mode semantic networks as shown in Fig. 4(a). The probabilities are shown close to the tip of each edge. We formally define the first mode Semantic Network as follows:

Definition 4 [Similarity based Network]. Let $X = \{x_1,...x_n\}$ be the set of artifacts, where each $x_i \in X$ has a feature vector $f_i = \{f_{i1}, ..., f_{im}\}$ then a first mode Similarity based Network $N^{sn}(V^{sn}, E^{sn})$ is a directed graph where V^{sn} is a set of nodes and E^{sn} is a set of edges, such that $V^{sn} \subseteq X$ and $|V^{sn}| \leq |X|$, and each edge links two relevant artifacts $<v_i, v_j>$ and has a probability score $w(v_i, v_j)$ where $0 < w(v_i, v_j) \leq 1$.

Rule Enforced Network

The automatically created first mode networks reflect similarity based on the feature vectors of each artifact but they do not include any additional semantic information. For example, there could be strong relevance between two nodes representing files from different projects, but because some attributes in the feature vectors (e.g. the name) are completely different, the Jaccard similarity coefficients may not rank them similar enough to create an edge between them. Such semantic knowledge is usually captured in the minds of experienced users, and it can be described in terms of *semantic rules* that explicitly identify connectivity between two nodes in the semantic

---

[3] A Jaccard similarity coefficient (Jaccard index) measures the similarity of sets and is defined as the size of the intersection divided by the size of the union of the sample sets.

network. This is required in two scenarios first the two nodes that were not deemed to be similar according to similarity measures (although they are similar indeed), second, there may be a situation where two nodes have a high similarity as per the similarity measures but have a low similarity. We define a semantic rule as follows:

Definition 5 [Semantic Rule]. *Given two artifacts x_p and x_q a semantic rule r is defined as r: x_p, x_q, w(xp , xq) where w(xp , xq) is the probability score associating the two artifacts.*

When these semantic rules are enforced, they add edges connecting nodes on the first mode semantic network(s), thus, they augment the network. The probabilities on the new edges are also calculated and the result is the second mode semantic network as shown in Fig. 2(b), which we refer to as *Rule enforced Network*. The new edges are depicted as dashed arrows. When multiple experts with similar roles create the same rule connecting two edges, we add a link to the network having as relevance score the average probability of all occurrences of the rule. When experts with different roles create new rules it is possible that these rules would expand the network in completely different directions. In such cases, we do not try to consolidate these rules into a single network, but we create separate networks each one specific to a role. We formally define this second mode Semantic Network.

Definition 6 [Rule enforced Network]. *Given a first mode Semantic Network $N^{sn}(V^{sn}, E^{sn})$, where V^{sn} is a set of nodes and E^{sn} is a set of edges in N^{sn}, and a set of semantic rules R, a second mode rule enforced Semantic Network $N^{re}(V^{re}, E^{re})$ is a directed graph where V^{re} is a set of nodes and E^{re} is a set of edges such that $V^{re} \subseteq X$, $|V^{re}| \leq |X|$ and $|V^{re}| \geq |V^{sn}|$, and each edge links two relevant artifacts $<v_i, v_j>$ and has a probability score $w(v_i, v_j)$ where $0 < w(v_i, v_j) \leq 1$.*

The probability scores encompass the similarity between the features of each artifact and the semantic rules enforced on the network. Such a network contains knowledge of multiple people, e.g., an individual programmer of TSAFE may not know the relationships of artifacts in FMS, and vice versa, but a software architect may know that TSAFE is related to FMS, although the software architect may not know in detail the relationships between the artifacts within each system. However, based on the relevance scores between neighboring nodes in the network, we can infer the relevance between any two nodes (as far apart as FMS 5 is to TSAFE II DM – see Fig. 1). Thus, one can discover more semantically related collective information compared to individual knowledge. If a semantic rule links two nodes that are already connected in the previously created similarity network, the semantic rule link replaces the similarity link (the expert's opinion supersedes the feature based similarity).

Algorithm for Automatic Semantic Network Generation

Once we have the feature vectors we then use the Jaccard coefficient to quantify the similarity among the feature vectors of the artifacts. We use the Jaccard coefficient since it does not give importance to a positive dissimilarity of features (marked as 0-0 in bits identifying that there is no similarity between two features that do not match) but gives importance to a positive match (1-1) and positive mis-match(1-0). We outline the approach to identifying the similarity of the feature vectors in Algorithm 2. The complexity of the algorithm is $O(N^2|U|)$ where N is the number of artifacts and $|U|$ is the size of the Universal Vector.

```
Algorithm 2 The Similarity based Network generation
algorithm
Require: Set of artifacts X where each x_i ∈ X is
associated with a set of attributes f_i = {f_i1,.. ,f_im}
Ensure: Similarity based Network N^sn(V^sn, E^sn) where V^sn
is a set of nodes and E^sn is a set of edges, each edge
links two relevant artifacts < v_i, v_j > and has a
probability score w(v_i, v_j)
 1: jc=0
 2: deg=0
 3: for i = 1 to |X| - 1 do
 4:    for j = i + 1 to |X| do
 5:       jc_ij = jc_ji = JC(f_i, f_j)
 6:          deg_i = deg_i + jc_ij
 7:          deg_j = deg_j + jc_ji
 8:    end for
 9: end for
10: for p = 1 to n do
11:    for q = 1 to n do
12:        w(xp , xq)  ←  jc_pq/deg_p
13:        if w(xp , xq) < W_threshold then
14:            { w(xp , xq)  = 0}
15:        end if
16:    end for
17: end for
```

4.3 Using Context

We store context in relational tables. One table stores user context, with columns user ID, project ID, role of user, programming language, etc. A second table stores application context, including project ID, functionalities, etc. A third table stores environmental context, including user ID, operating system, organization name, etc. After these tables have been initialized, we create a mapping table to map information in these tables to conditions on attribute-value pairs over the artifacts. For example, if the user's role is developer, we map it to the condition: Type = Code ∨ Type = Requirement as a developer needs to read both code and requirements.

We can then combine all context information of a user into a single filtering condition. This condition is the conjunction of all conditions mapped from the context information of a user. For example, a user's filtering condition may be:

(Type = Code ∨ Type = Requirement) ∧ (Programming Language = JAVA) ∧ (Project = Flight Control) ∧ (Operating System = LINUX)

At run time, this condition is used to filter the artifacts returned by the full-text search and semantic network. This step checks the attribute-value pairs of a returned artifact, and if any of those attributes in the artifact appears in the filtering condition, the value of that attribute will be checked against the filtering condition. If the value violates the condition, the artifact will be pruned. For example, if an artifact with Programming Language = C++ is returned, this artifact violates the above filtering condition and will be pruned.

Note that if an attribute of an artifact does not appear in the filtering condition, no check will be done and the artifact will remain in the result. For example, if programming language is not specified in an artifact (e.g., when the artifact is a design document), then this artifact will not be pruned based on the condition on programming languages.

4.4 Transformation and Composition Functions

It is quite intriguing to evaluate the effect of applying context during the different phases of the query lifecycle. For example, is it better to apply context before using semantic networks or after? Can we apply context both before and after using the semantic network? Questions like this might affect greatly the artifacts that will be retrieved and we investigate answers to these questions in this Section.

Each user query submitted to our system undergoes a series of transformations as it passes through its various phases and completes its cycle though our system (Section 4.1). During each of these different phases a transformation function is applied to a specific input available in the current phase, and produces a specific output applicable to the next phase. For example, extracted keywords of the initial user query are used as input to a function f_{MAS}, which conducts a Metadata Artifacts Search (MAS) and produces as its output a result containing artifacts R_A. Formally,

Definition 7 [Metadata Artifacts Search Function]. *Assume that Q_A is a set of keywords of a user query, and A is the domain of all artifacts. The function f_{MAS} is the Metadata Artifacts Search function which takes as input Q_A and produces as output a set of artifacts R_A .*

$$f_{MAS}(Q_A) = R_A \quad (alternatively \ Q_A \xrightarrow{f_{MAS}} R_A), \ where \ R_A \subset A.$$

In a similar fashion we define two more transformation functions: f_{SN} and f_C which apply the semantic network techniques and the context techniques respectively. Therefore we have:

Definition 8 [Semantic Network Transformation Function]. *The function f_{SN} applies the input R_A through a semantic network and produces as output a set of related arti-*

facts R_{SN} . $f_{SN}(R_A) = R_{SN}$ *(alternatively* $R_A \xrightarrow{f_{SN}} R_{SN}$), *where* $R_A, R_{SN} \subset A.$

Definition 9 [Context Transformation Function]. *The function f_C filters the input R_{SN} utilizing the appropriate context $C(U_j, T)$, and produces as output a set of filtered*

artifacts R_C . $f_C(R_{SN}) = R_C$ *(alternatively* $R_{SN} \xrightarrow{f_C} R_C$), *where* $R_{SN}, R_C \subset A.$

Definition 10 [Lifecycle Composition Function]. *A lifecycle composition function L of a user query in our system is a composition of the transformation functions f_{MAS}, f_{SN}, and f_C defined as* $L : f_C (f_{SN},(f_{MAS}(Q_A)))= R_C$. *Alternatively,* L
$: f_C \circ (f_{SN} \circ f_{MAS}) : Q_A \xrightarrow{f_{MAS}} R_A \xrightarrow{f_{SN}} R_{SN} \xrightarrow{f_C} R_C$ *where* R_A, R_{SN}, $R_C \subset A$.

We have used the above transformation functions in a specific order to compute the final result R_C as a composition of functions: $f_C \circ (f_{SN} \circ f_{MAS})$. Nevertheless, there are different ways that we can order the transformation functions and create a different composition. For example, $L_1 : f_{SN} \circ (f_C \circ f_{MAS})$ is another composition where the context function f_C is applied before the semantic network function f_{SN}. It is interesting to examine whether we obtain the same results depending on the order of the transformation functions in the function composition. In general, the composition of functions is not a commutative operation, i.e., $f_C \circ (f_{SN} \circ f_{MAS}) \neq f_{SN} \circ (f_C \circ f_{MAS})$. In practice, we can apply the transformation functions in different orders depending on how we want the process to take place, we can even apply the same transformation function multiple times. For example, it makes sense to apply the context function f_C before *and* after the semantic net function f_{SN}, having a new lifecycle composition function $L_2 : f_C \circ (f_{SN} \circ (f_C \circ f_{MAS}))$. We discuss the different options (L, L_1, and L_2) in our next Section where we describe our experiments.

5 Experiments

We first describe the setup of our experiments in section 5.1. In Section 5.2 we present our results for the automatic creation of the semantic network. Next we discuss the experiments with context and without context and in Section 5.3 we present the results. We use recall and precision as our basic measures according to the definitions of [73] and [6]. We also describe our prototype system in the Appendix.

5.1 Setup of Experiments

We used two test-beds of two software projects each:
   (1) The *Tactical Separation Assisted Flight Environment* (TSAFE) and the *Flight Management System* (FMS). These two software projects are based on a specification for Automated Air Traffic Control by NASA [21], implemented by MIT [19] and turned into a test-bed at Fraunhofer Center, Maryland [44]. This test-bed makes a good fit for the proposed research for two reasons. First, it contains two parallel threads of implementations of similar functionality. Second, historical design information exists for all variants and versions of TSAFE and FMS. There are as many as 38 different versions of each project, making the total number of artifacts more than 250, not counting the source code class files. The different versions of TSAFE and FMS are related, making reuse possible but not straightforward. Valuable design information can be retrieved; however, the different versions and amount of existing

data makes finding such design information difficult using the current full text search system.

(2) The second test bed consists of information from two different software projects; we selected 82 artifacts from DMGroup1 and 76 artifacts from LosGroup3. These two projects implemented similar functionality. We asked three domain experts to create an initial semantic network to use it as a baseline on these test beds. The relationship between two files in different projects (DMGroup1 and LosGroup3) receives a weight of 1 if these two files are deemed similar or a weight of 0 if they are dissimilar to each other. The metadata of each artifact includes the name of the artifact, the type of the artifact (requirement document, code, etc.), programming language, impact analysis (the impact of a specific software change to the various phases of software maintenance) , design pattern (a blueprint that can be applied to provide a solution to a commonly occurring problem), etc.

5.2 Automatic Semantic Network Generation

Creation of Feature Vectors

To create the feature vectors we used the second test bed with a set of 158 Java files from two different projects (82 files from DMGroup1 and 76 files from LosGroup3) as an input to a Java program. We used this test bed since we wanted to specifically evaluate the similarity among files across different projects. These files are compared with each other based on 4 main characteristics: java import statements (150), package names (120), class names (120) and method names (almost 300). A universal feature vector is automatically generated with a set of 680 attributes identifying characteristics which are unique across these 158 files. Next we compare each of these characteristics in every other file from two different projects to find whether they are similar or not. This similarity is captured in a similarity matrix which maps the similarity of each file with all other files across different projects. If two files from different projects are similar based on a Jaccard similarity coefficient and our weight computation then we mark the matrix location with a 1 otherwise with a 0. The similarity threshold for this task was set to 0.8. Different threshold values produce different results as described in Section 5.3.1. Source and edges along with weights are stored in an Oracle database which is subsequently used to build the tree structure.

Evaluation and Validation of the Automatically Created Network

The domain experts review the files from two different projects and label the similarity weights as 0 or 1. If they find similar files they give the weight 1, if the files are not similar they label them with a weight 0. Their concept of similarity is purely based on the manual evaluation of the artifacts and no specific features are considered. For the evaluation of the automatic network creation we consider this similarity provided by the domain experts as our labeled data where the domain expert provides a weight to the pairs of artifacts. Since the domain experts view is absolute numeric value of 0 or 1 we devised a method to check whether we did find similar files using our approach. We compared one artifact from one project with all the other artifacts in a different project and the one which has maximum similarity weight based on our approach was checked against the one provided by the domain experts as having the maximum similarity weight of 1.

	Probability from Rule Enforced Network		
Probability from Domain Experts		**Max similarity**	**No similarity**
	Max Similarity	144 (TP)	3 (FN)
	No similarity	10 (FP)	0 (TN)

$$\text{Accuracy} \quad = \quad \frac{TP \ + \ TN}{TP \ + \ TN \ + \ FP \ + \ FN} \quad \text{.........} \quad (1)$$

Fig. 5. Performance evaluation using Class labels

Based on this we validate against the labeled data and find the Accuracy of our method. Using the values from table in Fig. 5 we can compute the accuracy as shown in Equation 1. Thus, Accuracy = (145 + 3)/ (145+10+0+3) = 148/158 = 0.93*100 = 93%. From a set of 158 files taken from two projects (DMGroup1 and LOS Group3). Domain Experts found 148 files highly similar out of these 158 and our approach found 145 similar out of 158. Max Similarity is 145 i.e. artifacts which were found to be highly similar by our approach and the domain experts. Out of 158, there are 10 false positives where the domain experts found no similarity but our approach found some similarity, In addition we found 3 files which were highly similar which the domain experts did not identify.

5.3   Experiments with Context

For this set of experiments we used the data from the TSAFE/FMS test-bed. We collected eight queries from the domain experts. For each query, we also created eight different contexts by assuming a certain type of user (user context), a certain type of project (application context), a certain type of programming language (user context and/or application context), and a certain type of platform (environment context). Thus there are altogether 64 combinations of queries and contexts. The domain experts provided us with the correct answers to those queries. We compare the precision and recall of three search algorithms:

1. Using the normal full-text search algorithm without semantic network or context. We used Oracle's full-text search feature for this algorithm (referred to as No-Network in Fig. 6-9)
2. Using semantic network but not context (referred to as Network in Fig. 6-9)
3. Using both semantic network and context (referred to as Network+Context in Fig. 6-9)

5.3.1   Results
An important parameter in our approach is the threshold t for relevance score in the semantic network. We experimented using the default and also a varying threshold. Fig. 6 and 7 report the average recall and precision of all three algorithms using the default setting $t = 0.8$.

The x-axis identifies the queries, while the y-axis presents the value of recall (Fig. 6), and precision (Fig. 7), both averaged over the eight different contexts for each query. The results show that using a semantic network produces a much higher recall than not using a semantic network (see Fig. 6). This is expected because the semantic network returns artifacts that may not contain searched keywords, but are related to the artifacts containing those keywords.

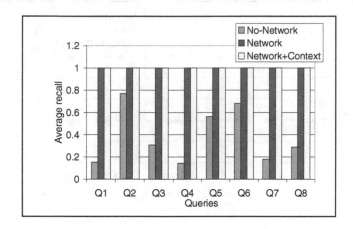

Fig. 6. Recall when threshold = 0.8

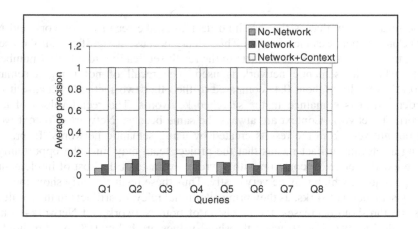

Fig. 7. Precision when threshold = 0.8

The results also show that the use of context increases precision because the context information is used to filter out results not relevant to the user (Fig. 7). In general, using both the semantic network and context leads to higher precision and recall for all eight queries (the recall and precision values at 1 occur due to the relatively small size of the data set).

Next we varied the relevance score threshold t in the semantic network. Fig. 8 and 9 report the average recall and precision over the eight queries when t varies from 1 down to 0.2. Note that No-Network does not use semantic network and it is interpreted as t being fixed at 1. Thus, the recall and precision of No-Network do not change with t.

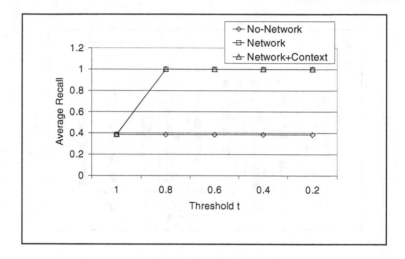

Fig. 8. Average recall with varying threshold

The results show that as the threshold decreases, the recall of Network and Network+Context increases (see Fig. 8). This occurs because it signifies that the user is willing to accept less relevant artifacts in the result set, leading to a higher number of results, when the semantic network is used. The recall of not using a semantic network is very low (about 0.4) compared to the other two methods because it only considers artifacts contained in the searched keywords. The recall values of using Network or Network+Context are always the same because Network+Context would filter out answers from the results created by using semantic networks. In practice, missing a relevant artifact means that the project team may miss the opportunity of reusing existing code; or come up with a wrong estimate of the cost of implementing a change request, which may be very costly. Thus these results clearly show the value of using semantic networks, as they bring additional relevant artifacts in the result set.

As the threshold decreases, the precision of both Network and Network+Context starts to decline (see Fig. 9) when threshold values are below 0.8. As threshold decreases below 1.0 but is still quite high (say, 0.8), artifacts which are very closely related to the answers in the full-text search are returned, and are considered as correct answers; thus, the precision remains high. However, as the threshold further decreases, artifacts that are not very closely related are returned. Thus, the precision starts to decline. This suggests that using a relatively high threshold (we use 0.8) would ensure both high precision and recall. Of course, if recall is very important (e.g., the cost of missing a relevant artifact is very high) a lower threshold can be used to ensure high recall, but with possibly lower precision.

The results also show that the use of context and semantic network always leads to higher precision than using semantic network alone, because context helps filter out irrelevant answers. Using Network+Context also has higher precision than No-Network over a wide range of threshold values (actually for all the threshold values we tested), and with a much higher recall as shown in the previous figures. It clearly displays the benefits of using a combined approach of semantic networks and context.

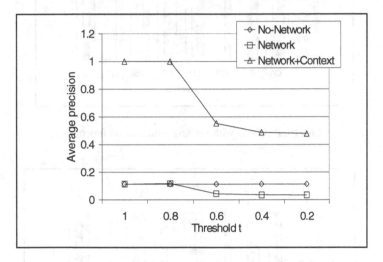

Fig. 9. Average precision with varying threshold

Experimenting with Different Lifecycle Composition Functions

The experiments we just described correspond to the lifecycle composition function L (see Section 4.5). We also performed another set of experiments using the alternate lifecycle composition functions L_1 (with context applied only before the semantic network $f_{SN} \circ (f_C \circ f_{MAS})$) and L_2 (with context applied before and after the semantic network $f_C \circ (f_{SN} \circ (f_C \circ f_{MAS}))$) and compared the results with those of L (with context applied only after the semantic network $f_C \circ (f_{SN} \circ f_{MAS})$). Fig. 10 and 11 report the precision and recall of these three composition functions.

The results show that L (using context after the semantic network) and L_2 (using context-based filtering both before and after using semantic network) produced the same results (both recall and precision) for six out of the eight queries. Query Q3 and Q7 were exceptions. For those two queries, when we applied context before the semantic network, it did not return any answer, resulting in the lowest recall and precision. The reason was that the direct hits (the results after full text search but without the semantic network) were actually "out-of-context". However, these direct hits were related to the correct answers that were in "in-context". Thus using L_2 the system filtered out the direct hits and did not return any correct answer. On the other hand, using L the system still used all the direct hits to find relevant artifacts through the semantic network, thus correct answers were still returned.

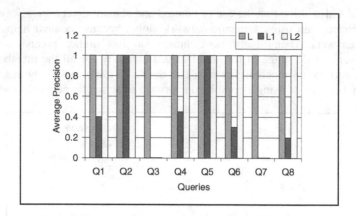

Fig. 10. Precision with varying composition functions

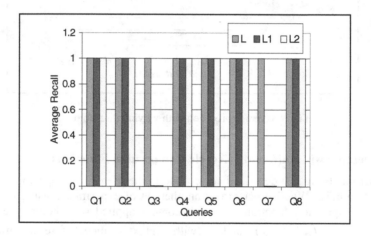

Fig. 11. Recall with varying composition functions

Both L_1 and L_2 apply context based filtering before the semantic network. However, L_2 applies context again after using the semantic network. Since using context after the semantic network does not eliminate any artifacts in the correct answer (i.e., matching the context), L_1 and L_2 have the same recall. The results also show that L_1 (using context-based filtering before semantic network) leads to lower precision than both L_2 and L. This is expected as the use of the semantic network augments the results with semantically related artifacts. But checking the context filtering condition before the use of the semantic network does not guarantee that the augmented results are "in context". For example, one of the contexts precludes source code for managers; still L_1 returns source code related to requirement documents which are in the correct answer. This also exemplifies the property of non-commutativity of the lifecycle composition function, $f_C \circ (f_{SN} \circ f_{MAS}) \neq f_{SN} \circ (f_C \circ f_{MAS})$.

In general, if the contexts transformation function f_C is used just as a filtering condition, then it is advisable to apply f_C after the semantic network transformation function f_{SN} has produced results. The reason is that this process will return artifacts that are connected to those initial artifacts that match the query, but do not match the context. Thus, in such cases the default composition function L should be used. Of course, if the contexts are used in other ways, e.g., to expand the query results or to modify a possible ranking function for displaying results, then it may make sense to use context transformation function f_C before the semantic network transformation function f_{SN}.

User Studies

We conducted two studies regarding the searching behavior of users. First, we carried out an experiment about search tools' performance [49], which confirmed our main hypothesis that retrieving information is difficult, especially for subjects unfamiliar in the domain of the search. Concepts, acronyms, and company lingo were information that most such subjects lacked in order to find the relevant information. In contrast, subjects with some company experience did not experience these problems. Based on these results, we developed and evaluated a prototype search tool that automatically manipulates the search query adding synonyms, acronyms, and abbreviations, increasing the relevance of the search results substantially.

Secondly, we performed another study [25], and we surveyed members of two small IT organizations, one in the US and another one in Germany regarding their search behavior. The results showed that more than 80% of the subjects used only one to four search keywords for their queries.

6 Conclusions

We created a set of tools and technologies which use metadata to assist software engineers in their search for software artifacts. Semantic networks capture the semantic relationships between software artifacts; these networks help return additional artifacts that are semantically relevant to the search, which would not have been included in the original search results using traditional database techniques. We provided an automated way to create the similarity based semantic network and described two algorithms towards its creation. Once the semantic network is built, it can be enhanced with semantic rules; subsequent user queries take advantage of the relationships that are represented in it. This technique produces an augmented result set of the user query, relevant to the original search, thus improving the recall of the search results. However, this augmented and relevant artifact set, may not be tailored to the appropriate contexts of the particular user. Therefore, we employ techniques to filter out "out-of-context" results, and return only "in-context" artifacts pertaining to the user. As a result, the precision is also improved.

We applied our techniques in a software engineering environment with software engineering projects. We performed experiments on real life software projects with the help of domain experts, measuring precision and recall, by comparing full-text, semantic network only, and a combined use of semantic networks and context methods. The results demonstrated that our metadata techniques are promising and they produce a better recall and precision results compared to the other methods.

In the future, we plan to investigate under which circumstances it would be better to use different orderings of transformation functions, by creating various lifecycle composition functions. In addition, during the automatic creation of the semantic network, the size of the universal vector may explode quickly as each artifact may potentially contribute new features to the universal vector U. Thus, to control the high dimensionality of U we will need to assign weights to the features which we defer to future work. Additionally in the current work we use a Jaccard similarity coefficient; however, in the future we plan to investigate other similarity coefficients such as Matching, Tanimoto, Cosine and Dice coefficients [34, 40] and compare them to Jaccard similarity coefficient.

Acknowledgements. This work was partially supported by an award from the US National Science Foundation (SGER 0738898) and by a grant from Northrop-Grumman Corporation.

References

1. Aamodt, A., Plaza, E.: Case-based reasoning: Foundational Issues, Methodological Variations, and System Approaches. Artificial Intelligence Communications 7(1), 39–59 (1994)
2. Agrawal, R., Rantzau, R., Terzi, E.: Context-sensitive ranking. In: ACM SIGMOD International Conference on Management of data, Chicago, IL, USA, pp. 383–394 (2006)
3. Aleman-Meza, B., Halaschek-Wiener, C., Arpinar, I.B., Ramakrishnan, C., Sheth, A.P.: Ranking Complex Relationships on the Semantic Web. IEEE Internet Computing, 37–44 (2005)
4. Antoniou, G., Hermelen, F.v.: A Semantic Web Primer, p. 238. The MIT Press, Cambridge (2004)
5. Anvik, J., Hiew, L., Murphy, G.C.: Who should fix this bug? In: International Conference on Software Engineering (2006)
6. Baeza-Yates, R.A., Ribeiro-Neto, B.A.: Modern Information Retrieval. ACM Press / Addison-Wesley (1999)
7. Basili, V.R., Rombach, H.D.: Support for Comprehensive Reuse. IEEE Software Engineering Journal 6(5), 303–316 (1991)
8. Bazire, M., Brezillon, P.: Understanding Context Before Using It. In: Dey, A.K., Kokinov, B., Leake, D.B., Turner, R. (eds.) CONTEXT 2005. LNCS (LNAI), vol. 3554, pp. 29–40. Springer, Heidelberg (2005)
9. Bergmann, R., Göker, M.: Developing Industrial Case-Based Reasoning Applications Using the INRECA Methodology. In: Workshop at the International Joint Conference on Artificial Intelligence, IJCAI - Automating the Construction of Case Based Reasoners, Stockholm (1999)
10. Berners-Lee, T., Hendler, J., Lassila, O.: The Semantic Web. Scientific American 284(5), 34–43 (2001)
11. Bohner, S.A., Arnold, R.S.: Software Change Impact Analysis. IEEE Computer Society Press, Los Alamitos (1996)
12. Bouquet, P., Giunchiglia, F., van Harmelen, F., Serafini, L., Stuckenschmidt, H.: C-OWL: Contextualizing ontologies. In: Fensel, D., Sycara, K., Mylopoulos, J. (eds.) ISWC 2003. LNCS, vol. 2870, pp. 164–179. Springer, Heidelberg (2003)

13. Brezillon, P.: Context in Human Machine Problem Solving: A Survey. LAFORIA (1996)
14. Canfora, G., Cerulo, L.: Impact analysis by mining software and change request reposito-
    ries. In: International Software Metrics Symposium, METRICS 2005 (2005)
15. Canfora, G., Cerulo, L., Penta, M.D.: Relating software interventions through IR tech-
    niques. In: International Conference on Software Management (2006)
16. Chen, Z., Gangopadhyay, A., Holden, S., Karabatis, G., McGuire, M.: Semantic Integra-
    tion of Government Data for Water Quality Management. Journal of Government Informa-
    tion Quarterly – Special Issue on Information Integration 24(4), 716–735 (2007)
17. Chen, Z., Gangopadhyay, A., Karabatis, G., McGuire, M., Welty, C.: Semantic Integration
    and Knowledge Discovery for Environmental Research. Journal of Database Manage-
    ment 18(1), 43–67 (2007)
18. Cubranic, D., Murphy, G.C.: Automatic bug triage using text categorization. In: Interna-
    tional Conference on Software Engineering & Knowledge Engineering, Banff, Alberta,
    Canada, pp. 92–97 (2004)
19. Dennis, G.: TSAFE: Building a Trusted Computing Base for Air Traffic Control Software.
    MIT, Cambridge (2003)
20. Diligenti, M., Coetzee, F., Lawrence, S., Giles, C.L., Gori, M.: Focused Crawling Using
    Context Graphs. In: 26th International Conference on Very Large Data Bases, Cairo,
    Egypt, pp. 527–534 (2000)
21. Erzberger, H.: Transforming the NAS: The Next Generation Air Traffic Control System.
    In: 24th International Congress of the Aeronautical Sciences (2004)
22. Fankhauser, P., Kracker, M., Neuhold, E.J.: Semantic vs. Structural Resemblance of
    Classes. SIGMOD Record 20(4), 59–63 (1991)
23. Farquhar, A., Dappert, A., Fikes, R., Pratt, W.: Integrating Information Sources using Con-
    text Logic. In: AAAI Spring Symposium on Information Gathering from Distributed Het-
    erogeneous Environments (1995)
24. Fayyad, U., Piatetsky-Shapiro, G., Smyth, P.: From Data Mining to Knowledge Discovery:
    An overview. In: Advances in Knowledge Discovery and Data Mining. AAAI/MIT Press
    (1996)
25. Feldmann, R.L., Rech, J., Wenzler, A.J.: Experience Retrieval in LSOs: Do you find what
    you are looking for? In: 8th International Workshop on Learning Software Organizations
    (LSO 2006), Rio de Janeiro, Brazil (2006)
26. Fellbaum, C.: WordNet: An Electronic Lexical Database, p. 423. MIT Press, Cambridge
    (1998)
27. Finkelstein, L., Gabrilovich, E., Matias, Y., Rivlin, E., Solan, Z., Wolfman, G., Ruppin, E.:
    Placing Search in Context: the Concept Revisited. In: WWW, pp. 406–414 (2001)
28. Georgakopoulos, D., Karabatis, G., Gantimahapatruni, S.: Specification and Management
    of Interdependent Data in Operational Systems and Data Warehouses. Distributed and Par-
    allel Databases 5(2), 121–166 (1997)
29. Glass, R.: Agile Versus Traditional: Make Love, Not War. Cutter IT Journal, 12–18 (2001)
30. Goh, C.H., Bressan, S., Madnick, S., Siegel, M.: Context Interchange: New Features and
    Formalisms for the Intelligent Integration of Information. ACM Transactions on Informa-
    tion Systems 17(3), 270–293 (1999)
31. Gong, L., Riecken, D.: Constraining Model-Based Reasoning Using Contexts. In: IEEE In-
    ternational Conference on Web Intelligence, WI 2003 (2003)
32. Guha, R., McCool, R., Fikes, R.: Contexts for the Semantic Web. In: McIlraith, S.A.,
    Plexousakis, D., van Harmelen, F. (eds.) ISWC 2004. LNCS, vol. 3298, pp. 32–46.
    Springer, Heidelberg (2004)

33. Haney, F.M.: Module Connection Analysis - A Tool for Scheduling Software Debugging Activites. In: AFIPS Joint Computer Conference, pp. 173–179 (1972)
34. Haranczyk, M., Holliday, J.: Comparison of Similarity Coefficients for Clustering and Compound Selection. Journal of Chemical Information and Modeling 48(3), 498–508 (2008)
35. Joachims, T.: Optimizing search engines using clickthrough data. In: ACM SIGKDD international conference on Knowledge discovery and data mining, Edmonton, Alberta, Canada, pp. 133–142 (2002)
36. Johnson, J.H.: Micro Projects Cause Constant Change. In: 2nd International Conference on eXtreme Programming and Flexible Processes in Software Engineering, pp. 132–135 (2001)
37. Karabatis, G.: Using Context in Semantic Data Integration. Journal of Interoperability in Business Information Systems 1(3), 9–21 (2006)
38. Karabatis, G., Rusinkiewicz, M., Sheth, A.: Interdependent Database Systems. In: Management of Heterogeneous and Autonomous Database Systems, pp. 217–252. Morgan-Kaufmann, San Francisco (1999)
39. Kashyap, V., Sheth, A.: Semantic and schematic similarities between database objects: a context-based approach. The VLDB Journal 5(4), 276–304 (1996)
40. Kaufman, L., Rousseeuw, P.J.: Finding Groups In Data: An Introduction To Cluster Analysis. Wiley-Interscience, Hoboken (2005)
41. Kraft, R., Maghoul, F., Chang, C.C.: Y!Q: Contextual Search at the Point of Inspiration. In: ACM International Conference on Information and Knowledge Management, Bremen, Germany, pp. 816–823 (2005)
42. Lieberman, H., Selker, T.: Out of context: Computer Systems that adapt to, and learn from, context. IBM Systems Journal 39(3&4), 617–632 (2000)
43. Lindvall, M., Feldmann, R.L., Karabatis, G., Chen, Z., Janeja, V.P.: Searching for Relevant Software Change Artifacts using Semantic Networks. In: 24th Annual ACM Symposium on Applied Computing SAC 2009, Honolulu, Hawaii, U.S.A., pp. 496–500 (2009)
44. Lindvall, M., Rus, I., Shull, F., Zelkowitz, M.V., Donzelli, P., Memon, A., Basili, V.R., Costa, P., Tvedt, R.T., Hochstein, L., Asgari, S., Ackermann, C., Pech, D.: An Evolutionary Testbed for Software Technology Evaluation. Innovations in Systems and Software Engineering - a NASA Journal 1(1), 3–11 (2005)
45. Lindvall, M., Sandahl, K.: How Well do Experienced Software Developers Predict Software Change? Journal of Systems and Software 43(1), 19–27 (1998)
46. Liu, H., Singh, P.: ConceptNet: A Practical Commonsense Reasoning Toolkit. BT Technology Journal 22(4), 211–226 (2004)
47. Lormans, M., Deursen, A.v.: Can LSI help Reconstructing Requirements Traceability in Design and Test? In: Conference on Software Maintenance and Reengineering, CSMR 2006 (2006)
48. Lucca, G.D., Penta, M.D., Gradara, S.: An approach to classify software maintenance requests. In: International Conference on Software Maintenance, Los Alamitos, CA (2002)
49. Lydie, Y.T.M.: Context-Based Information Retrieval -User Problems and Benefits of Potential Solutions, Technical Report. FC-MD (2006)
50. Marcus, A., Maletic, J.I.: Recovering Documentation-to-Source-Code Traceability Links using Latent Semantic Indexing. In: 25th International Conference on Software Engineering, ICSE 2003 (2003)
51. Masterman, M.: Semantic message detection for machine translation, using an interlingua. NPL, 438–475 (1961)

52. McCarthy, J.: Notes on formalizing context. In: International Joint Conference on Artificial Intelligence (IJCAI), Chambéry, France, pp. 555–560 (1993)
53. McGuinness, D.L., Harmelen, F.v.: OWL Web Ontology Language Overview W3C (2004), http://www.w3.org/TR/owl-features/
54. Mockus, A., Herbsleb, J.D.: Expertise browser: a quantitative approach to identifying expertise. In: International Conference on Software Engineering, New York, NY, pp. 503–512 (2002)
55. Mylopoulos, J., Cohen, P., Borgida, A., Sugar, L.: Semantic Networks and the Generation of Context. In: International Joint Conference on Artificial Intelligence, Tiblisi, Georgia, pp. 134–142 (1975)
56. Ostertag, E., Hendler, J., Prieto-Diaz, R., Braun, C.: Computing similarity in a reuse library system: An AI-based approach. ACM Transactions on Software Engineering and Methodology 1(3), 205–228 (1992)
57. Ostertag, E.J.: A Classification System for Software Reuse, Ph.D. Dissertation. University of Maryland (1992)
58. Pearl, J.: Probabilistic Reasoning in Intelligent Systems: Networks of Plausible Inference. Morgan Kaufmann, San Francisco (1988)
59. Pearl, J.: Causality: Models, Reasoning, and Inference. Cambridge University Press, Cambridge (2000)
60. Pomerol, J.-C., Brezillon, P.: Dynamics between Contextual Knowledge and Proceduralized Context. In: Bouquet, P., Serafini, L., Brézillon, P., Benercetti, M., Castellani, F. (eds.) CONTEXT 1999. LNCS (LNAI), vol. 1688, pp. 284–295. Springer, Heidelberg (1999)
61. Prieto-Diaz, R.: Implementing faceted classification for software reuse. Communications of the ACM 34(5), 89–97 (1991)
62. Prieto-Diaz, R.: Status report: Software reusability. IEEE Software 10(3), 61–66 (1993)
63. Prieto-Diaz, R., Freeman, P.: Classifying software for reusability. IEEE Software 4(1), 6–16 (1987)
64. Rice, J.A.: Mathematical Statistics and Data Analysis. Duxbury Press (1994)
65. Rusinkiewicz, M., Sheth, A., Karabatis, G.: Specifying Interdatabase Dependencies in a Multidatabase Environment. IEEE Computer 24(12), 46–53 (1991)
66. Shen, X., Tan, B., Zhai, C.: Context-Sensitive Information Retrieval using Implicit Feedback. In: 28th international ACM SIGIR conference on Research and development in information retrieval, Salvador, Brazil, pp. 43–50 (2005)
67. Sheth, A., Aleman-Meza, B., Arpinar, I.B., Bertram, C., Warke, Y., Ramakrishanan, C., Halaschek, C., Anyanwu, K., Avant, D., Arpinar, F.S., Kochut, K.: Semantic Association Identification and Knowledge Discovery for National Security Applications. Journal of Database Management 16(1) (2004)
68. Sheth, A., Karabatis, G.: Multidatabase Interdependencies in Industry. In: ACM SIGMOD International Conference on Management of Data, Washington, DC (1993)
69. Sowa, J.F.: Semantic Networks, http://www.jfsowa.com/pubs/semnet.htm
70. Sowa, J.F.: Semantic Networks. In: Shapiro, S.C. (ed.) Encyclopedia of Artificial Intelligence, pp. 1493–1511. Wiley, New York (1992)
71. Sowa, J.F.: Laws, Facts, and Contexts: Foundations for Multimodal Reasoning. In: Hendricks, V.F., Jorgensen, K.F., Pedersen, S.A. (eds.) Knowledge Contributors, pp. 145–184. Kluwer Academic Publishers, Dordrecht (2003)
72. Sugiyama, K., Hatano, K., Yoshikawa, M.: Adaptive web search based on user profile constructed without any effort from users. In: WWW (2004)

73. Tan, P.-N., Steinbach, M., Kumar, V.: Introduction to Data Mining. Addison-Wesley, Reading (2006)
74. TopicMap: XML Topic Maps (XTM) 1.0, http://www.topicmaps.org/xtm/
75. W3C: Semantic Web (2001), http://www.w3.org/2001/sw/
76. Wache, H., Stuckenschmidt, H.: Practical Context Transformation for Information System Interoperability. In: Akman, V., Bouquet, P., Thomason, R.H., Young, R.A. (eds.) CONTEXT 2001. LNCS (LNAI), vol. 2116, pp. 367–380. Springer, Heidelberg (2001)
77. Wu, H., Siegel, M., Ablay, S.: Sensor Fusion for Context Understanding. In: 19th IEEE Instrument and Measurement Technology Conference, Anchorage, AK, USA (2002)
78. Zimmermann, A., Lorenz, A., Oppermann, R.: An operational definition of context. In: Kokinov, B., Richardson, D.C., Roth-Berghofer, T.R., Vieu, L. (eds.) CONTEXT 2007. LNCS (LNAI), vol. 4635, pp. 558–571. Springer, Heidelberg (2007)

Applying Fuzzy DLs in the Extraction of Image Semantics

Stamatia Dasiopoulou[1], Ioannis Kompatsiaris[1], and Michael G. Strintzis[1,2]

[1] Centre for Research and Technology Hellas, Informatics and Telematics Institute,
Thessaloniki, Greece
[2] Information Processing Laboratory, Electrical and Computer Engineering
Department, Aristotle University of Thessaloniki, Greece
{dasiop,ikom,michael}@iti.gr

Abstract. Statistical learning approaches, bounded mainly to knowledge related to perceptual manifestations of semantics, fall short to adequately utilise the meaning and logical connotations pertaining to the extracted image semantics. Instigated by the Semantic Web, ontologies have appealed to a significant share of synergistic approaches towards the combined use of statistical learning and explicit semantics. While the relevant literature tends to disregard the uncertainty involved, and treats the extracted image descriptions as coherent, two valued propositions, this paper explores reasoning under uncertainty towards a more accurate and pragmatic handling of the underlying semantics. Using fuzzy DLs, the proposed reasoning framework captures the vagueness of the extracted image descriptions and accomplishes their semantic interpretation, while resolving inconsistencies rising from contradictory descriptions. To evaluate the proposed reasoning framework, an experimental implementation using the fuzzyDL Description Logic reasoner has been carried out. Experiments in the domain of outdoor images illustrate the added value, while outlining challenges to be further addressed.

1 Introduction

Semantic image analysis has challenged researchers for decades in the quest for generalisable approaches to alleviate the so called *semantic gap* [1,2,3,4], i.e. the lack of correspondence between the descriptions that can be automatically extracted from visual content and the respective meaning a human would attach. Towards this goal, statistical learning approaches have attracted increased interest in the last couples of years, as they provide powerful and effective means to discover, capture, and manage, complex associations between perceptual features (i.e. attributes of visual manifestations such as colour and texture) and semantic concepts. Support Vector Machines (SVMs) [5] and Bayesian Networks (BNs) [6] constitute popular examples, and have been espoused in numerous approaches targeting the extraction of image semantics [7,8,9,10,11,12].

Although promising results have been reported, the attained performance varies significantly with respect to the number of concepts addressed and the

S. Spaccapietra, L. Delcambre (Eds.): Journal on Data Semantics XIV, LNCS 5880, pp. 105–132, 2009.

considered image data sets as well. The observed variability relates to core challenges in computer vision, including perceptual similarities of semantically distinct concepts and perceptual variations in the possible manifestations of a single concept [13,14], much as to limitations inherent in the assumption that semantics can be rendered in a visual fashion. As a result of the aforementioned, the learnt associations do not necessarily reflect the intended semantics, viz. the associations implicitly targeted when providing the corresponding training examples, leading often to incomplete and conflicting classifications. Indicatively, among the observations presented in [15], the weapon classifier proves to be more efficient when querying for palm trees, while when querying for instances of fire and flames, the soccer classifier provides the highest performance.

Acknowledging the value of statistical learning techniques, yet aware of their weaknesses, approaches towards the synergistic utilisation of explicit semantics have become the subject of systematic research[1]. The Semantic Web (SW) incentive influenced to a large extent the choice of the investigated representation formalisms, favouring the use of SW languages [16,17] and of the closely related Description Logics (DLs) [18,19]. As a result, a number of multimedia ontologies [20,21,22,23] have been proposed to represent perceptual features and to enable linking with domain specific ontologies, in order to formalise the transition from low-level features to semantic entities [24,25,26,27]. In addition, domain ontologies, tailored to the analysis viewpoint as well, have been proposed in order to acquire interpretations of higher abstraction through reasoning over automatically extracted descriptions [28,29,30,31,32].

However, the effects entailing from the poor utilisation of semantics in statistical learning, namely the uncertainty inherent in the extracted descriptions and the semantic inconsistencies issuing from conflicting descriptions, tend to be overlooked. Specifically, the extracted classifications are commonly treated as crisp assertions, neglecting significant information regarding the plausibility of the acquired descriptions. Furthermore, in the majority of cases, the extracted descriptions are assumed to be semantically coherent. As a result, the use of explicit conceptual models and reasoning are rendered mainly as means to acquire abstract and complex descriptions by exploiting logical associations between the extracted descriptions, such as the inference of a person instance by reasoning over instances of face and body in a certain configuration. Evidently though, both aforementioned assumptions are rather weak and hardly correspond to the pragmatics of the problem at hand.

Aiming to enhance the utilisation of semantics and alleviate part of the aforementioned effects in the accuracy and completeness of descriptions that are extracted by means of machine learning approaches, we present in this paper a reasoning framework that utilises fuzzy DLs semantics in order to interpret the

---

[1] Indicatively, besides individual research activities, this pursuit has been the principal objective in a substantial number of European projects including aceMedia (http://www.acemedia.org/aceMedia), K-Space (http://kspace.qmul.net), BOEMIE (http://www.boemie.org/), X-Media (http://www.x-media-project.org/), MESH (http://www.mesh-ip.eu/), SALERO (http://www.salero.eu/), etc.

output of the classifiers into a semantically consistent interpretation. The use of DLs allows us to formally capture the semantics underlying the concepts of interest, while the fuzzy extensions provide the means to model the vagueness encompassed in the extracted classifications. Furthermore, extending on earlier investigations [33], the presented framework supports the explicit representation of the constituent image regions, allowing, as explained in the sequel, the more effective utilisation of the underlying semantics. The contribution of the proposed reasoning framework can be summarised in the following.

- The uncertainty of the descriptions made available through the application of learning based approaches is formally handled and taken into consideration in the interpretation of the descriptions' semantics.
- The inconsistencies resulting from conflicting descriptions, due to the aforementioned limitations in the learning of associations between perceptual features and corresponding semantics, are identified and resolved.
- Besides formally grounding the acquisition of the most plausible interpretations in the presence of multiple possible interpretations, the proposed fuzzy DLs based reasoning framework supports the identification of image regions where concepts, missed in the initial descriptions, may be present.

The rest of the paper is organised as follows. Sections 2 and 3 outline the reasons that motivated our investigation and the particular issues involved in the application of formal reasoning in semantic image analysis. Section 4 presents the proposed reasoning framework architecture and its constituent reasoning tasks, while Section 5 elaborates the implementation details. Section 6 presents the evaluation of the proposed framework and the experiences drawn. Relevant initiatives are presented in Section 7, while Section 8 summarises the paper and outlines future research directions.

2   Motivation

Statistical concept classifiers exhibit highly variable performance, yet support generic learning for a substantial number of concepts [15,34,35,36]. As demonstrated in a recent study [37], satisfactory retrieval can be achieved, even when the detection accuracy is low, provided that sufficiently many concepts are used, as long as these concepts can be related to one another in some reasonable way. In addition, the conducted experiments reveal that when there exist semantic associations between the addressed concepts, then these concepts can serve as an intermediate layer to enhance the reliability of the extracted semantic image descriptions. The conducted experiments consider semantic video descriptions, yet the results can be easily generalised for the case of image descriptions, since the examined concept classifiers address notions detected per video frames, i.e. without the use of temporal information.

The observations drawn by the aforementioned study regarding the potential of incorporating semantics, do not outline a new direction; approaches following the knowledge-directed paradigm have been reported since the early 70s, while they boomed in the 80s and the early 90s in accordance with the respective advances in the field of Artificial Intelligence (AI) [38,39,40]. Yet, [37] stresses the greater potential that the recent advances in statistical concept classifiers conduce regarding the utilisation of explicit knowledge and reasoning as the means to alleviate the limitations related to the discriminative capacity of perceptual features with respect to the intended semantics.

As already described though, the limitations related to the rather poor utilisation of semantics are intertwined with the uncertainty involved in extracting semantic descriptions from images. As such, the fundamental question of what constitutes the semantics of this uncertainty, emerges. The answer lies in the viewpoint adopted in learning regarding the stipulation of semantics in accordance to perceptual features. Approaches where concepts are detected on the grounds of perceptual similarity, imply a prototypical set of feature values that constitute a visual/perceptual definition of the concept. As the presence of a concept is determined based on the similarity of those values, concepts can be considered as fuzzy sets, where the similarity (distance) function serves the role of the membership function. Contrariwise, learning approaches that utilise concepts' co-occurrence and correlation, implement a probabilistic interpretation of the features to concepts transition. Support Vector Machines (SVMs) constitute a popular example of the former category, while Bayesian Nets and Hidden Markov models [41] fall in the latter.

Apparently, both types of uncertainty pertain to the extraction of image semantics, much more since they address complementary aspects. A classification indicating that a specific image region constitutes an instance of sea with a probability of 0.7, refers to the presence or not of sea; no information is provided about how blueish this sea region might be. A classifier though that assess an image region to belong to the sea concept with a degree of 0.7, quantifies the similarity of this region with what has been learned as the perceptual definition of sea. For further details on the different semantics of the two uncertainty types, the reader is referred to [42]. The investigation of a reasoning framework that considers both types of uncertainty is undoubtedly of particular interest. In this work though, we focus on the fuzzy perspective of the extracted semantic descriptions, since we consider it an significant starting point for the appropriate handling of semantic classification results, and a useful insight into the complementary role of probabilistic reasoning.

The aforementioned incentives, in combination with the limited support for handling uncertainty and inconsistency provided by the state of the art approaches in the utilisation of explicit semantics, designated the selection of fuzzy DLs as the investigated knowledge representation. The logic grounded semantics ensure conceptual transparency and well-defined reasoning mechanisms, while maintaining a strong connection to the Semantic Web community. In addition, the fuzzy extensions allow to formally capture the imprecision in the from of vagueness that pertains to learning approaches based on perceptual similarity, such as SVMs. In combination with the particular expressivity requirements described in the following Sections, the aforementioned have been the main reasons for preferring fuzzy DLs over some other logic based formalism, such as fuzzy first order logic, or fuzzy rules.

3   Fuzzy DLs in Semantic Image Analysis: Specifications and Requirements

Fuzzy DLs extend the model theoretic semantics of classical DLs [18] to fuzzy sets [43,44] and account for a significant share of the literature studying the representation of imprecise information [45,46,47,48,49,50]. Standardisation initiatives, such as the W3C Uncertainty Reasoning for the World Wide Web Incubator Group, which recently released the final report on reasoning under uncertainty in the Semantic Web[2], outline further the significance of handling imprecise knowledge in real world applications.

The semantics of a fuzzy DL language are given by a fuzzy interpretation $I = (\Delta^I, \cdot^I)$, where Δ^I is an non-empty set of objects comprising the domain of interpretation, and $\cdot^I$ a fuzzy interpretation function, which assigns each individual a to an element $a^I \in \Delta^I$, each concept name A to a membership function $A^I : \Delta^I \to [0,1]$, and each role name R to a membership function $R^I : \Delta^I \times \Delta^I \to [0,1]$ [47,48]. Table 1, illustrates the standard interpretation of typical DL constructors.

Table 1. Fuzzy interpretation of DL constructors following Zadeh semantics [49]

$$\top^I = 1$$
$$\bot^I = 0$$
$$(\neg\, C)^I = 1\text{-}C^I(d)$$
$$(C \sqcap D)^I = \min\{C^I(d), D^I(d)\}$$
$$(C \sqcup D)^I = \max\{C^I(d), D^I(d)\}$$
$$(\forall\, R.C)^I = \inf_{d' \in \Delta} max\{1 - R^I(d, d'), C^I(d')\}$$
$$(\exists\, R.C)^I = \sup_{d' \in \Delta} min\{R^I(d, d'), C^I(d')\}$$

A fuzzy knowledge base consists of a TBox defined by a finite set of fuzzy concept inclusion and equality axioms, and an ABox defined respectively as a finite set of *fuzzy assertions*. A fuzzy assertion [47] is of the form $a : C \bowtie n$ and $(a, b) : R \bowtie n$, where $\bowtie$ stands for $\geq, >, \leq$, and $<$. Intuitively a fuzzy assertion of the form $a : C \geq n$ means that the membership degree of the individual a to the concept C is at least equal to n. The standard reasoning services (e.g. instance checking, satisfiability, subsumption etc.) are adapted analogously. For example, concept satisfiability with respect to C requires the existence of an interpretation under which there will be an individual belonging to C with a degree $n \in (0, 1]$.

Using fuzzy DLs as the knowledge representation language for the semantic interpretation of descriptions provided by statistical concept classifiers, renders the

---

[2] http://www.w3.org/2005/Incubator/urw3/XGR-urw3-20080331/

Fig. 1. Example outdoor image and segmentation mask

$(im : Rockyside) \geq 0.50$ $(im : Countryside\_buildings) \geq 0.47$
$(im : Roadside) \geq 0.48$ $(im : Forest) \geq 0.65$
$(im : Seaside) \geq 0.46$

$(r0 : Building) \geq 0.68$ $(r0 : Trunk) \geq 0.54$
$(r1 : Sky) \geq 0.70$ $(r1 : Person) \geq 0.59$
$(r2 : Building) \geq 0.66$ $(r2 : Trunk) \geq 0.58$
$(r3 : Vegetation) \geq 0.56$ $(r3 : Rock) \geq 0.51$
$(r4 : Building) \geq 0.66$ $(r4 : Spectators) \geq 0.54$
$(r5 : Trunk) \geq 0.55$ $(r5 : Building) \geq 0.53$
$(r6 : Building) \geq 0.61$ $(r6 : Board) \geq 0.51$
$(r7 : Building) \geq 0.60$ $(r7 : Board) \geq 0.52$
$(r8 : Tree) \geq 0.56$ $(r8 : Grass) \geq 0.55$

Fig. 2. Scene and object level classifications results for the example image of Fig. 1 using SVM-based concept classifiers

available classifications into fuzzy assertions and the available domain knowledge into corresponding terminological axioms. Figures 1 and 2, illustrate an example outdoors image, its segmentation mask and the extracted classifications in the form of fuzzy DLs assertions.

As illustrated, an image may be asserted to belong to multiple scene level concepts, not necessarily semantically related; similarly, a region may belong to multiple object level concepts. This is not unusual and accounts for two equally common situations met in the extraction of semantic image descriptions. First, the use of multiple classifiers for a single concept in order to benefit from multiple sources of information, and second, classification errors that result in false positive responses for semantically contradictory concepts. For readability, we consider at most two instances per region.

Assuming the TBox of Table 2 and going through the respective assertions, one notices that there exist semantic discrepancies with respect to the extracted scene level descriptions, since according to axioms $1 - 6$ only one of them can be true, as well as between the scene level descriptions and the object level ones.

Table 2. Example TBox extract for the domain of outdoor images

axiom 1: Forest $\sqsubseteq$ Landscape $\sqcap \neg$ (Countryside_buildings $\sqcup$ Roadside)
axiom 2: Roadside $\sqsubseteq$ Landscape $\sqcap \neg$ (Forest $\sqcup$ Countryside_buildings)
axiom 3: Countryside_buildings $\sqsubseteq$ Landscape $\sqcap \neg$ (Forest $\sqcup$ Roadside)
axiom 4: Landscape $\sqsubseteq$ Outdoors $\sqcap \neg$ (Rockyside $\sqcup$ Seaside)
axiom 5: Seaside $\sqsubseteq$ Outdoors $\sqcap \neg$ (Landscape $\sqcup$ Rockyside)
axiom 6: Countryside_buildings $\sqsubseteq \exists$ contains.Building $\sqcup \exists$ contains.Grass
axiom 7: Countryside_buildings $\sqcap \exists$contains.(Spectators $\sqcup$ Board $\sqcap$ Rock) $\sqsubseteq \bot$

Contradictions may be straightforward, such as in the case of (im:Roadside) $\geq$ 0.52 and (im:Forest) $\geq$ 0.65, or implicit such as in the case of (im:Forest) $\geq$ 0.65 and (r0:Building) $\geq$ 0.68, where through inference the latter assertion entails that (im:Countryside_buildings) $\geq$ 0.68. Furthermore, the identification of inconsistencies depends on the scene level concept used as a reference. Assuming for example that the Forest scene description is valid, all region assertions referring to the Building, Spectators and Board concepts entail inconsistency. Assuming though that the Countryside_buildings scene description is valid, inconsistencies are raised by regions assertions referring to the concepts Spectators, Board and Rock instead.

Consequently, in order to reach a coherent interpretation, the possible alternative scene interpretations need to be identified and subsequently assessed with respect to their plausibility. This means that for all possibly satisfiable scene concepts, that is for all scene concepts for which a model exists when conflicts, the corresponding degrees of membership need to be computed in order to provide a measure for their plausibility. Due to the logical relations between the object and scene level concepts, the degree to which an image belongs to a scene concept does not necessarily equal the degree provided by the respective scene concept classifier. Hence, in our current example, the satisfiable, and thereby plausible, scene descriptions are (im:Countryside_buildings), (im:Forest), and (im:Rockyside); the corresponding minimum degrees are 0.68, 0.65 and 0.51.

Once the most plausible scene description is determined, the next step is to ensure that the object level descriptions are not introducing semantic conflicts. As in the case of scene level descriptions, the identification, tracking and resolving of such inconsistencies is intertwined with the semantics as defined in the TBox axioms. In the considered example, the identification and resolving of inconsistencies is rather straightforward, since all inconsistent assertions refer to atomic concepts (i.e. Spectators, Rock, and Board). In the presence of an axiom such as $Person \sqcap Bench \sqsubseteq Spectators$ though, the inconsistency could be resolved in multiple ways, namely by removing all Person instances, all Bench instances, or all instances of both classes. Selecting among the different alternatives needs to take into account cost criteria encompassing the assertions' degrees of confidence in order to retain the available plausibility information.

The final step towards a more complete image description is to compensate for missing assertions and enrich the existing descriptions by means of entailment. Missing assertions refer to scene or object level descriptions that are entailed by the computed scene interpretation, yet failed to be detected by the applied concept classifiers. As elaborated in subsection 4.3, for the case of object level descriptions, the proposed framework allows not only to recover the missing assertions, but also to acquire suggestions regarding which of the input region instances could be a possible match for the missing object level descriptions. Enrichment on the other hand covers those cases where scene and object level descriptions of higher abstraction can be inferred from the available ones. In the running example, where Countryside_buildings is designated as the most plausible scene description, the presence of at least one region belonging to the Building concept and one region belonging to the Grass concept is entailed, each with a degree ≥ 0.68. As a result, the degrees of the region assertions referring to the concept Building are updated, and so is the assertion concerning region $r8$, which now becomes most plausible from the initially extracted one referring to the concept Tree. Furthermore, due to axioms 3 and 4, the image is also asserted as an instance of Landscape and Outdoors.

We note the significance of the existential ($\exists$) and the disjunction ($\sqcup$) constructors for the aforementioned tasks. The existential quantification allows to handle cases where the initial descriptions are incomplete, due to segmentation fault or to erroneous classification, while the union constructor allows to represent and reason over the alternative scene interpretations in order to assess their satisfiability. Considering rule formalisms instead, we would lack the possibility to express existential quantification or use disjunction in the head of rules to so as to state the entailment of multiple possible alternatives. Using fuzzy first order logic, the latter would not pose a problem, yet we would be unable to infer the existence of regions corresponding to concepts failed to be detected by the classifiers, as described in detail in the following Section.

Finally, it is important to stress that the TBox aims to capture generic knowledge reflecting the logical associations issuing from the semantics of the concepts at hand, rather than data set specific conceptualisations, as the latter would risk false implications when invoked on classifications over data sets with differing attributes. For example, an axiom such as $\exists contains.Tree \sqcap \exists contains.Trunk \sqsubseteq Forest$ may be representative for forest scenes for a given data set, yet in the general case it could lead to biased inferences, as Tree and Trunk instances can be as well found in many other scene descriptions. Restricting the included axioms to strict domain semantics modelling, the domain TBox can ensure that the extracted descriptions are in compliance with the semantics of the concepts they refer to.

4   Fuzzy DLs-Based Reasoning Framework for Semantic Image Analysis

Figure 3 depicts the proposed reasoning framework for managing the tasks outlined in the previous Section. As shown, the semantic interpretation of the

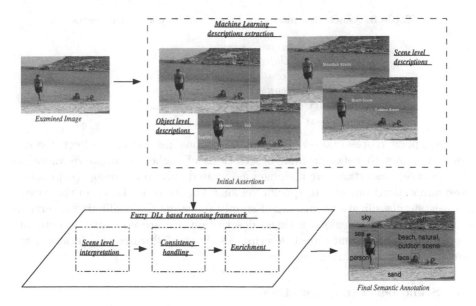

Fig. 3. Proposed fuzzy DLs based reasoning framework

descriptions that are acquired through statistical learning is performed in three steps. First, the most plausible description at scene level is determined. Next, the inconsistencies in the initial descriptions are resolved with respect to the previously computed scene level interpretation. As, more than one plausible interpretations may exist, during this step the different alternatives are ranked with respect to plausibility criteria. The set of assertions with the highest ranking is finally passed to the last step, where by means of logical entailment, assertions pertaining to complex or missing descriptions are made explicit.

Extending the conceptual framework of [33], in the current approach, we exploit localisation information for descriptions at object level. This is accomplished by explicitly representing the region which is associated with the extracted object concept descriptions, using the model of Table 3. According to it, an image and its constituent regions are associated through the role contains, while disjoint axioms make explicit the discrimination between scene and object level concepts, as well as between an image and its regions. As illustrated, there is no restriction on the number of descriptions at scene (respectively object) level that may be assigned to an image (region).

Towards a more conceptually accurate model, the first axiom would need to be revised as Image ⊑ ∃depicts.SceneConcept, so as to capture the fact that the Image concept includes those objects in the domain of interpretation that are associated with a SceneConcept instance through a depicts property. Similarly for the third axiom which would become Region ⊑ ∃depicts.ObjectConcept. As in the examined context though, both models would ensue the same inferences, we preferred the model of Table 3 to avoid unnecessary complexity.

Table 3. Annotation model for image and its constituent parts

$$Image \sqsubseteq SceneConcept$$
$$Region \sqsubseteq ObjectConcept$$
$$Image \equiv \exists contains.Region$$
$$Image \sqcap Region \sqsubseteq \bot$$
$$SceneConcept \sqcap ObjectConcept \sqsubseteq \bot$$

The explicit representation of the image regions has a twofold effect. On one hand, it allows to generate final descriptions of higher informative value, as object concept assertions are no longer associated only to the image (supporting loose annotation) but also to specific regions. On the other hand, in the case of inconsistent classifications, instead of simply removing the conflicting assertions and ending up with regions with no description, we are now able to compute suggestions of consistent object descriptions. In the following, the details of the three reasoning tasks are given.

4.1 Scene Level Interpretation

The alternative scene interpretations constitute the possible models regarding the interpretation of an image, and as a consequence for each $SceneConcept_i$ it suffices to find one model such that $SceneConcept_i \neq \emptyset$, instead of requiring $SceneConcept_i \neq \emptyset$ for all models. Towards this end, and due to the logical associations between concepts referring at the object level and concepts referring at the scene level, all assertions need to be taken into account in order to check the satisfiability of the alternative scene descriptions. As a consequence, all disjointness axioms in which scene level concepts are participating need to be removed before checking for satisfiability, as otherwise possible inconsistencies would reasoning and would prevent the effective utilisation of all information carried in the available extracted descriptions.

The scene interpretation procedure, summarised in Table 4, consists in the following steps. First, all disjoint axioms are removed and the TBox is revised with respect to the currently examined concept $SceneConcept_i$ so as to avoid conflicts when an image is inferred to belong both to $SceneConcept_i$ and its complement $\neg SceneConcept_i$. To accomplish this, the disjointness axioms are revised so as to entail an instance of $notSceneConcept_i$ instead of triggering an inconsistency. Considering the example TBox of 2, and for $SceneConcept_i$ equal to Rockyside, the presence of a region (r_i:Building $\geq d_i$), with $d_i \geq 0.5$ due to axiom 5 would entail (im:$\neg$ Rockyside ≥ 0.5), rendering the available classifications inconsistent. Revising axiom 5 as $Land- scape \sqsubseteq notRockyside$ though, the ABox remains consistent, and the assertion (im:notRockyside $\geq d_i$) is obtained instead.

Next, the satisfiability of each scene level description is checked considering all initial assertions besides the told scene level assertions that refer to a scene concept other than the currently examined one. Each scene concept $SceneConcept_i$, for which $notSceneConcept_i$ is not satisfiable constitutes a possible interpretation.

Table 4. Scene level interpretation

Scene Level Interpretation Algorithm
Input: scene level concepts hierarchy H_{SC}, input assertions A
Output: glb for all satisfiable scene level concepts

```
 1:   for all hierarchy levels L_i starting from the root
 2:      for all scene level concepts SC_j ∈ L_i
 3:         if ∃ satisfiable subsumee of SC_j or i==0{
 4:            revise disjoint axioms adding nonSC _j
 5:            check ¬ SC_j satisfiabity
 6:              if ¬ SC_j not satisfiable{
 7:                  remove assertions inconsistent to SC_j
 8:                  update A and compute glb(SC_j)
 9:              }
10:         }
11:   rank scene level concepts wrt glb
```

Thereby, compared with [33] the checks required to determine the most plausible scene descriptions, are reduced. To further improve the efficiency, the checking of the scene concepts satisfiability utilises the subsumption relations between the scene concepts. Thus, if a concept $SceneConcept_i$ is computed to be unsatisfiable, we skip the checks of all concepts subsuming it. For all satisfiable scene level concepts, the respective greater lower bound (glb) values are computed, by following the inconsistency handling methodology described in the sequel. Subsequently, the glb values are ranked and the scene concept with the highest one is selected as the most plausible scene description.

4.2 Inconsistency Handling

Having computed the scene level concepts that constitute possible interpretations, the next step is to obtain for each of them the most optimistic interpretation, in order to assess the most plausible one. Towards this end, for each satisfiable scene concept, the inconsistencies with respect to the input object concept assertions need to be identified and resolved. Following a similar procedure to the one described above, the TBox is revised so that the disjointness axioms involving the examined scene concept and object level ones, instead of causing an inconsistency, entail an instance of a correspondingly introduced $nonObjectConcept_i$. As more than one object level concepts $ObjectConcept_i$ may give rise to inconsistencies, a conjunctive expression is formed including the respective $nonObjectConcept_i$ concepts and the generic $nonObjectConcept$ concept is defined as it subsumer. To resolve the inconsistencies, we employ the procedure described in the following, until no instance of $nonObjectConcept$ with glb greater or equal than 0 exists.

Table 5. Expansion rules for computing the alternative sets of consistent assertions

⊓-rule	if $(a : C_1 \sqcap C_2) \in L(\mathrm{x})$
	then $L(\mathrm{y}) = L(\mathrm{x}) \setminus \{(a : C_1)\}$ and
	$L(\mathrm{z}) = L(\mathrm{x}) \setminus \{(a : C_2)\}$ and
	$L(\mathrm{w}) = L(\mathrm{x}) \setminus \{(a : C_1 \sqcap C_2)\}$
⊔-rule	if $(a : C_1 \sqcup C_2) \in L(\mathrm{x})$
	then $L(\mathrm{x}) = L(\mathrm{x}) \setminus \{(a : C_1 \sqcup C_2)\}$

where $C_i \longrightarrow \mathrm{A} \mid \mathrm{C} \sqcap \mathrm{D} \mid \exists \mathrm{R.D}$

First, we address inconsistencies incurred directly by told descriptions. This translates into checking whether there exist asserted individuals belonging to $ObjectConcept_i$ concepts such that $ObjectConcept_i \sqsubseteq nonObjectConcept_i$. The handling of such assertions is rather straightforward and consists in their removal. Addressing asserted individuals first, prunes the search space during the subsequent tracking of the inferences that lead to an inconsistency. Next, we consider assertions referring to complex concepts, i.e. concepts for which the left hand side of the axioms in which they participate is an expression rather than an atomic concept. Contrary to the previous case, we now need to analyse the involved axioms in order to track the asserted descriptions that cause the inconsistency. Furthermore, these axioms determine which of the descriptions should be removed so as to reach a consistent interpretation. To accomplish this, we build on relevant works for resolving unsatisfiable DL ontologies [51,52], and employ a reversed tableaux expansion procedure, summarised in Table 5.

The main difference with respect to the relevant literature is that in our application framework, we consider solely the removal of assertions, rather than the removal or weakening of terminological axioms. The expansion procedure starts having as root node the $(im : nonObjectConcept \geq d_i)$ assertion, where d_i the computed degree, and continues until no expansion rule can be applied. As illustrated, in the case of inconsistencies caused by axioms involving the conjunction of concepts, there are multiple ways to resolve the inconsistency and reach a consistent interpretation. Specifically, there as many alternative interpretations as the sum of combinations $C(N, k)$, where N the number of conjuncts and $k = 1, .., N$. In order to choose among them, we rank the set of solutions according to the number of assertions that need to be removed and the average value of the corresponding degrees. Again, corresponding $nonObjectConcept_i$ definitions are added as in the case of scene concepts to avoid ending up with inconsistent ABoxes.

4.3   Enrichment

The final step considers the enrichment of the descriptions by means of typical fuzzy DLs entailment. Specifically, once the scene level interpretation is determined and all inconsistencies have been resolved, we end up with a semantically consistent subset of the input assertions. To render the inferred descriptions explicit, corresponding queries are formulated and the responses are added to the final image interpretation. Inferred descriptions may refer either to concepts not addressed by the available classifiers (the Landscape concept constitutes such a concept for the example considered in Figures 1,2), or to concepts for whom the corresponding classifiers failed to produced a positive response.

Extending the framework of [33], the explicit representation of the constituent image regions, allows to model object level descriptions as instances referring to specific regions of the image, rather than to the entire image though the indirect representation of regions in the form of $(im : \exists contains.ObjectConcept_i)$ assertions. Such modelling allows for additional benefits besides the enhancement of loose image descriptions[3]. Specifically, once the inconsistency handling task is completed, there might be regions for which all initial assertions have been removed. Exploiting the visual coherency between the initial assertions associated to these regions and the assertions identified as missing, we can infer possible suggestions regarding object level concepts that such regions may depict. Furthermore, based again on the visual coherency of the concepts addressed by the classifiers, additional suggestions for missing assertions can be inferred with respect to regions that already have an object level concept assigned to them.

In order to capture and model the visual coherency of the considered concepts, we utilise the confusion matrixes acquired during the training phase of the classifiers and extract axioms of the form $ObjectConcept_i \sqsubseteq ObjectConcept_j \sqcup ObjectConcept_{j+1} \sqcup \ldots$, where the concepts $ObjectConcept_{j+n}$ represent object concepts that tend to be misclassified as instances of $ObjectConcept_i$ under a given scene description. The main reason for adopting such an approach, is that regions depicting visually similar object concepts often happen to be falsely segmented as one. As illustrated in the evaluation Section 6, the purpose of such suggestions is to facilitate the interaction with a subsequent step of analysis, including possibly re-segmentation and the re-application of specific classifiers on selected regions.

5   Implementation

In the previous Section, we described the individual tasks comprising the proposed fuzzy DLs-based reasoning framework for the enhancement of semantic image interpretation. In the following, we examine the proposed reasoning framework from an implementation perspective. Since, each task utilises corresponding standard fuzzy DLs reasoning services in order to accomplish its goals, central role in the proposed reasoning framework holds the reasoning engine that realises these core fuzzy DLs inference services.

The choice regarding which specific implementation should be employed under the proposed framework was based on the existing available fuzzy DL reasoning

---

[3] Loose (weak) annotation refers to object level descriptions that are associated to the entire image rather than the specific image regions.

engines and the requirements posed with respect to expressivity power and interaction capabilities. The sequence of works by Straccia [47,48,53] and Stoilos et al.
[49,50,54] distill the advancements accomplished with respect to the formal definition of fuzzy extensions semantics and of corresponding reasoning algorithms.
Complementary to the theoretic foundations, respective reasoning engine implementations have been developed, namely the *Fuzzy Reasoning Engine*[4] (FiRE)
and the *fuzzyDL*[5].

FiRE [55] supports querying an f-*SHIN* knowledge base for satisfiability, consistency, subsumption, and entailment, under Zadeh semantics; general concept
inclusions, roles and datatype support are among the planned future extensions.
fuzzyDL [56] supports satisfiability, consistency, subsumption, and entailment
for the language fuzzy *SHIF*, further extended by concrete fuzzy concepts, i.e.
concept defined through an explicit fuzzy membership function, concept mod-
ifiers that allow to change the membership function of a fuzzy concept, and
functional datatypes attributes. The reasoner accepts three types of semantics
for the interpretation of conjunction, disjunction, complement and implication,
namely Zadeh semantics, Lukasiewicz, and crisp. Although both available reasoners support very high expressivity and provide support for the standard reasoning services of satisfiability, instance checking, disjointness and subsumption,
the factor that differentiates them is the handing of general concept inclusions.
As illustrated in Sections 2 and 3, handling general concept inclusions is crucial as it allows to model the existence of specific regions, thus specific object
level concept instances, which in turn imply corresponding scene concept instances. Otherwise, the object level instances would be reduced to scene concept
instances, and subsequently the region individuals would become tautological to
the respective image individual. Given the above considerations, we selected the
fuzzyDL reasoner.

Fig. 4 shows an abstract view of the proposed reasoning framework architecture regarding the interaction between the proposed reasoning framework
and the fuzzyDL reasoning engine. As illustrated, and already described in the
detailed presentation of the procedure comprising each of the three reasoning
tasks, the fuzzyDL engine provides the standard inference services required to
support the semantic interpretation of an image. The proposed reasoning framework coordinates the required inference services by designating the each time
considered TBox and ABox, performing appropriate translations to avoid inconsistencies and formulate respective queries so as to determine the conditions
for the subsequent processing steps. Hence, it serves as an external mechanism
that modularises and harmonises the interpretation into distinct subproblems
on which the fuzzyDL can be invoked.

In addition, the proposed framework provides support for subtasks addressing
the handling of semantics that cannot be invoked as distinct services of fuzzyDL.
One such example is the tracking and resolving of inconsistencies, where besides
the transformation of the TBox so that an inconsistency entails an instance of

---

[4] http://www.image.ece.ntua.gr/∼nsimou
[5] http://faure.isti.cnr.it/∼straccia/software/fuzzyDL/fuzzyDL.html

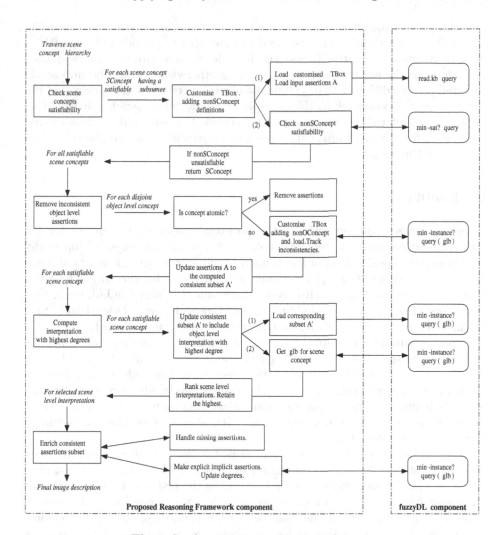

Fig. 4. Implementation architecture diagram

respectively introduced nonConcepts, the semantics of the axioms involved in the creation of an inconsistency are taken into account in order to compute the possible alternative solutions. We note that regarding the TBox revision, parts of the TBox translations, specifically the transformations during the satisfiability checks in the scene level interpretation task, have been manually performed, as the emphasis in the current experimental implementation is placed on assessing the feasibility of the proposed approach.

Another example is the computation of possible models with respect to a specific scene concept, given a TBox and the initial region assertions. In this case, first the TBox needs to be revised so that inconsistencies can be tracked and removed, and afterwards the possible models (i.e. the different configurations regarding the association of each region to one of the disjuncts representing the available classification results) need to be explicitly provided to fuzzyDL as distinct ABoxes. The latter is essential in order to compute the corresponding glb values for the examined scene concept, as otherwise it would be impossible to entail a value different than ≥ 0 unless all disjuncts per region addressed the same concept. Going back to the example of Figure 2 for instance, it would be impossible to infer the scene concept Countryside_buildings unless there existed a region r_i for which all alternative possible classification where the same one, i.e. that of Building.

6   Evaluation

In order to assess the potential of the proposed reasoning framework for the purpose of enhancing the semantic coherency and completeness of image descriptions, we experimented in the domain of outdoor images. The sets of scene (C_{scene}) and object (C_{object}) level concepts addressed by the employed SVM based classifiers are C_{scene}={Roadside, Rockyside, Countryside_buildings, Seaside, Forest} and C_{object}={Building, Roof, Grass, Foliage, Dried-Plant, Sky, Rock, Tree, Sea, Sand, Boat, Road, Ground, Person, Trunk, Wave}.

Figure 5 illustrates example images of the addressed scene level concepts. As illustrated, Seaside images refer to coastal and beach scenes, Rockyside ones include mountainous images with little vegetation, Roadside images refer to landscape scenes depicting parts of road, Countryside_buildings represent scenes where buildings are present yet not in an urban environment, and finally Forest images correspond to natural landscapes with abundant vegetation, including trees, foliage, trunks, etc.

From an initial set of 700 outdoor images, two sets of 350 images have been assembled: one served as the training set on which the learning of concept classifiers was performed, and the second served as the test set. Ground truth for all images has been manually generated at object and scene level. The manual annotation and training of the classifiers, both comprising quite cumbersome and resource intensive activities (corresponding to an average of two to three

Seaside   Rockyside   Roadside   Countryside   Forest
                                          buildings

Fig. 5. Example images of the supported scene level classifiers

Table 6. Example extract of the outdoor image domain TBox developed for evaluation purposes

> Countryside_buildings ⊑ ∃contains.Building ⊓ ∃contains.Grass
> Countryside_buildings ⊑ Landscape
> Grass ⊔ Tree ⊑ Foliage
> Rockyside ⊑ ∃contains.Rock
> Roadside ⊑ ∃contains.Road
> Roadside ⊑ Landscape
> ∃contains.Building ⊑ Countryside_buildings
> ∃contains.Sea ≡ Seaside
> Beach ≡ Seaside ⊓ ∃contains.Sand
> ∃contains.Sky ⊑ Outdoor
> Trunk ⊑ Tree
> Wave ⊑ Sea
> Boat ⊑ Sea
> Forest ⊓ (Roadside ⊔ Countryside_buildings) ⊑ ⊥
> Roadside ⊓ Countryside_buildings ⊑ ⊥
> Rockyside ⊓ (Seaside ⊔ Landscape)⊑ ⊥
> Landscape ⊑ ⊓ Outdoor ⊥
> Forest ⊓ ∃ contains.(Rock ⊔ Sea ⊔ Sand ⊔ Building ⊔ Road) ⊑ ⊥
> Rockyside ⊓ ∃ contains.(Sea ⊔ Sand ⊔ Building ⊔ Road) ⊑ ⊥

person months - for the number of concepts and images considered in the specific experiment), constitute efforts already spent for the purpose of training and assessing the performance of the employed SVM based classifiers [57]. Thus, the only extra resources required with respect to the proposed reasoning framework relate to the transformation of the already existing ground truth files and extracted descriptions to a format compliant to the one used by the proposed reasoning framework, which amounts to a negligible amount of effort. The reason for stressing this out, is to outline that the application of the proposed reasoning framework does not entail any additional resources with respect to annotation tasks.

In order to apply the proposed reasoning framework, a TBox that captures the semantics of the domain addressed by the available classifiers needs to be constructed. Table 6 illustrates an extract of the outdoor images TBox that has been developed for the carried out experimentation. It includes 25 concepts and one role. The included concepts comprise the scene and object level concepts supported by the classifiers, Landscape, Outdoor, and the generic SceneConcept and ObjectConcept concepts that are used to enforce that the discrimination between the two levels of concepts; the respective role is *contains*, which links an image to its constituent regions, and scene level concepts to object level ones.

Approximately fifty axioms, including the transformations required to avoid inconsistencies with respect to the alternative scene level interpretations, are used to capture the interrelations of the involved scene and object level concepts; this number increases further, when taking into account the additional axioms appended during the handling of inconsistencies. We note though, that not all axioms are loaded at once to fuzzyDL, since the proposed reasoning framework coordinates, as previously explained, the axioms and assertions over which the reasoning services of fuzzyDL are invoked. As a consequence, the complexity remains too low to incur performance concerns, and similar observations have resulted when experimenting with larger TBoxes, as long as the considered ABoxes remained similar in size.

Specifically, using a virtual Linux machine, running on XP Windows, with an Intel Core quad processor, requires about three hours and twenty minutes to process the complete test set. Individual image processing times, vary from two seconds to one minute and half, depending on the given assertions and the complexity of resolving the encountered inconsistencies. Actually, the inconsistency handling process, which computes the possible consistent alternatives by tracking the definitions involved, and the satisfiability and glb queries communicated to fuzzyDL are the most time consuming tasks. The average memory required per image is 26 MBs, of which only a small fragment, namely 1/100, is consumed by the proposed framework, the rest committed by the evoked fuzzyDL services. Both observations relate to the fact that the proposed reasoning framework addresses mostly the coordination of the input and queries to be communicated to the fuzzyDL than realising itself core inference services, with the exception of tracking inconsistencies.

In order to quantify the performance of the proposed approach, we compared the accuracy and completeness of the obtained image descriptions, with the descriptions provided by means of classification, as well as with the descriptions acquired when using the reasoning framework of our previous study [33]. The last allows for a first estimate on the added value of explicitly representing the individual image regions and the alternative object descriptions associated with them. As evaluation metrics, we adopted recall, precision and and F-measure, according to the following definitions.

- Precision (p): number of correct assertions extracted/inferred per concept divided by the number of assertions that were extracted/inferred for the given concept.
- Recall (r): number of correct assertions extracted/inferred per concept divided by the number of assertions referring to that concept that are present in the ground truth image descriptions.
- F-measure: $2 * p * r / (p + r)$.

Table 7 gives the performance of the classifiers, of the reasoning framework presented in [33], and of the currently proposed reasoning framework, for the case of scene level concepts. Compared to the performance of the classifiers, we note that the application of the proposed reasoning framework incurs a significant improvement. Going through the respective domain axioms, it is easy to correlate the extend of enhancement to the extent of semantic relations between object level concepts with scene level, particularly axioms that entail a scene level descriptions based on object level descriptions. Compared with the respective reasoning performance of [33], the explicit representation of the individual

Table 7. Evaluation of analysis and reasoning performance for scene level concepts

Concept	Analysis			Reasoning [33]			Reasoning		
	Recall	Precision	F-M	Recall	Precision	F-M	Recall	Precision	F-M
Rockyside	0.68	0.70	0.69	0.68	0.79	0.74	0.65	0.77	0.72
Seaside	0.85	0.67	0.75	0.86	0.72	0.78	0.79	0.75	0.78
Beach	-	-	-	0.45	0.76	0.57	0.45	0.76	0.57
Roadside	0.68	0.69	0.69	0.72	0.70	0.70	0.72	0.63	0.67
Forest	0.75	0.63	0.69	0.74	0.68	0.71	0.76	0.68	0.72
Countryside buildings	0.30	1.0	0.46	0.60	0.86	0.71	0.60	0.86	0.71
Landscape	0.75	0.71	0.	0.87	0.85	0.85	0.87	0.85	0.85
Outdoor	-	-	-	1.0	1.0	1.0	1.0	1.0	1.0

image regions and the corresponding object level assertions appears to have a rather negligible effect, as the slight improvement observed for concepts such as Roadside and Forest is counterbalanced by the slight deterioration with respect to the Rockyside and Seaside concepts.

Table 8 compares the respective performance for descriptions at object level. As shown, besides the Boat and Grass concepts, the application of the reasoning framework of [33] improves significantly the performance compared to the sole application of the classifiers. This is a direct consequence of the fact that the considered object level concepts are characterised by rich semantics with respect to the scene level concepts that constitute their context of appearance. The behaviour observed with respect to the Boat and Grass concepts relates to the risks entailed by a false scene level interpretation, which may incur in the case of very poor classification performance, in which case the input descriptions suggest interpretations other than the actual one. Going for example through the images for which Boat assertions where falsely removed, thus incurring the observed lowering in the recall rate, we noticed that the corresponding prevailing scene level assertions were not incompliance with the actual scene semantics.

Similar considerations emerge when analysing the not so remarkable effect of reasoning in the recall of scene level concepts such as Rockyside. Going through the images depicting rocky side scenes, yet failed to be recognised as such, we noticed that in all cases the classifiers had falsely detected another scene level concept instead, despite the fact that the instantiations of the Rock concept were successfully detected in the their majority. Adding an axiom such as ∃contains.Rock ⊑ Rockyside would seem a reasonable idea for improving performance on the grounds that the available axioms did seem to overlook this knowledge. However, as in the case of Boat, such an amendment would imbalance the trade off between what constitutes domain semantics and what is mere tuning to the peculiarities of a given data set. In the discussed case, this is easy to illustrate simply considering how often it is for rocks to appear in beach scenes.

The application of reasoning however, under the model proposed in this paper that considers the individual regions, entails an even higher effect on the

Table 8. Evaluation of analysis and reasoning performance for object level concepts

Concept	Analysis			Reasoning[33]			Reasoning		
	Recall	Precision	F-M	Recall	Precision	F-M	Recall	Precision	F-M
Building	0.54	0.69	0.60	0.62	0.86	0.72	0.62	0.64	0.63
Roof	0.33	0.54	0.41	0.33	0.75	0.46	0.43	0.63	0.52
Grass	0.49	0.42	0.45	0.30	0.52	0.38	0.83	0.56	0.67
Vegetation	0.48	0.84	0.61	0.86	0.86	0.86	0.80	0.49	0.61
Dried-Plant	0.07	0.11	0.08	0.07	0.13	0.10	0.12	0.33	0.18
Sky	0.95	0.93	0.94	0.95	0.93	0.94	0.96	0.92	0.94
Rock	0.65	0.45	0.53	0.69	0.70	0.69	0.57	0.57	0.57
Tree	0.49	0.52	0.51	0.56	0.47	0.51	0.83	0.46	0.59
Sand	0.02	0.10	0.03	0.57	0.45	0.50	0.57	0.45	0.50
Sea	0.69	0.60	0.64	0.85	0.69	0.76	0.75	0.69	0.72
Boat	0.41	0.71	0.52	0.33	0.66	0.44	0.44	0.57	0.5
Road	0.50	0.69	0.58	0.69	0.71	0.70	0.77	0.52	0.62
Ground	0.26	0.33	0.29	0.26	0.33	0.29	0.49	0.45	0.47
Person	0.75	0.51	0.61	0.75	0.51	0.61	0.86	0.45	0.61
Trunk	0.26	0.28	0.27	0.26	0.28	0.27	0.33	0.22	0.27
Wave	0.25	0.5	0.33	0.25	0.5	0.33	0.25	0.5	0.33

completeness and accuracy of the object level descriptions. This a direct consequence of the fact that instead of leaving a region without a corresponding assertion in the case the classification results prove to be inconsistent, probable suggestions are inferred that as illustrated incur further improvement. In order to obtain the values illustrated in the Table, we considered for each region the inference-based suggestion with the highest degree. As described in Section 4 though, more than one suggestions may be inferred for a given region, independently of whether this region has been subjected to inconsistent classification, aspiring to further assist in the identification of additional descriptions. As a result, in the case of a more interactive analysis and classification module, the proposed framework has the potential for an effectively higher enhancement.

Table 9 provides a rough assessment of the potential benefit such suggestions entail, by measuring the respective recall and precision values when all suggested additional descriptions are taken into account. As expected, the concepts that exhibit the higher potential for improvement are those for which once the scene description has been identified, their perceptual similarity with already detected concepts allows them to be associated to existing region assertions. As described, these suggestions have disjunctive semantics, i.e. they do not necessitate the presence of the suggested concept but rather identify the most plausible regions at which this concepts should be sought. Figure 6 provides an estimation of the number of regions that should be searched if no such information was available, i.e. when the only knowledge relates to the region assertions that have been missed during classification, and the respective number of regions when the suggestions provided by the proposed reasoning framework are used, provided

Table 9. Evaluation of reasoning for object level concepts including the inferred suggestions

Concept	Reasoning		
	Recall	Precision	F-M
Building	0.66	0.83	0.74
Roof	0.35	0.69	0.46
Grass	0.60	0.75	0.67
Vegetation	0.6	0.68	0.64
Dried-Plant	0.05	0.22	0.08
Sky	0.95	0.93	0.94
Rock	0.65	0.45	0.53
Tree	0.49	0.52	0.51
Sand	0.02	0.10	0.03
Sea	0.69	0.60	0.64
Boat	0.47	0.71	0.52
Road	0.64	0.78	0.70
Ground	0.27	0.31	0.28
Person	0.75	0.51	0.61
Trunk	0.41	0.33	0.37
Wave	0.25	0.5	0.33

that they are correct; otherwise, the searching for a missing object reduces to the same situation as in the former case. As illustrated when the inferred suggestion are taken into account, the number of regions that need be examined is reduced almost by half for concepts for which semantic and perceptual information is available.

Summing up the experiences and observations drawn from the conducted evaluation, we note that the utilisation of explicit semantics has a positive impact towards the semantic interpretation of image descriptions. The use of fuzzy DLs allows to handle formally the degrees of confidence that accompany the automatically extracted and utilise them both towards the identification of the most plausible interpretation as well as for resolving inconsistencies. The preservation of the degrees information in combination with the ensured semantic coherency of the resulting image descriptions, renders the proposed framework a useful contribution for semantic retrieval tasks that address multimedia content. Furthermore, since the presented framework makes no assumption with respect to the classifiers used to provide the initial classifications, it has the potential to be employed in any image retrieval scenario involving vague. descriptions.

Indicatively, practical cases where the proposed framework could be employed include applications such as the TRECVID[6] challenge, where among the addressed tasks is the extraction of high-level visual content descriptions using statistical learning. Within such context, the proposed reasoning framework could be used to alleviate inconsistent classifications and to enhance the completeness

---

[6] http://www-nlpir.nist.gov/projects/trecvid/

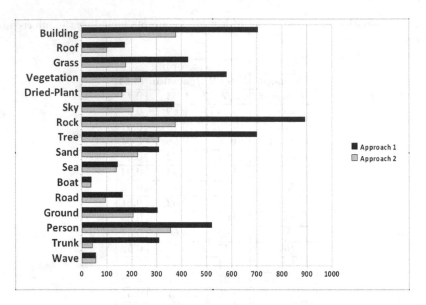

Fig. 6. Comparison of the number of regions that need to be examined for missing object level assertions, when no additional knowledge is available (approach 1) and when the inferred suggestions are taken into consideration (approach2)

of the final content descriptions. Thereby, the reliability of the descriptions is improved, while concepts that are not supported by the classifiers, but are semantically related, can be afforded, sparing the time and effort for building such classifiers. Another example application where the proposed framework could be used is the DL-Media [58] retrieval system in order to allow the ontological query service to perform over inconsistent image descriptions. In general, as exemplified in the motivating examples and the carried out evaluation, the proposed framework has the potential to enhance content descriptions, and by consequence the corresponding content management tasks, acquired by means of typical statistical learning techniques, when the underlying imprecision refers to vagueness.

7   Related Work

The majority of relevant literature considers the investigation of crisp DLs-based approaches. In the series of works presented in [28,59,29], crisp DLs are proposed for inferring descriptions modelled as logical aggregates. A probabilistic approach is described in the more recent one as a possible solution to the handling of the ambiguity introduced in visual analysis [29]. Although the proposed approach outlines an interesting direction, it lacks the technical details and evaluation that would establish the potential contribution; furthermore, considering probabilistic information, it addresses a different kind of uncertainty than what is presented in this paper.

In [31], DLs have been extended with a rule-based approach to realise abductive inference over crisp analysis assertions. Alternative consistent interpretations are computed by means of abduction and ranked using as criteria the number of new individuals that need to be introduced[7] and the number of assertions that need to be left out in order to reach a consistent interpretation. Examining the combined use of such an abductive reasoning framework with the proposed one fuzzy, could be interesting for investigating the effect in the ranking of alternative interpretations.

In [60] DLs are used to realise the interpretation of feature values pertaining to colour, texture and background knowledge to semantic objects. To this end a pseudo fuzzy algorithm is presented to reason over the calculated feature values with respect to the prototypical values constituting the definition of semantic objects. Additionally, topological knowledge is utilised to exclude inconsistent associations of semantic objects to given image segments. More specifically, in addition to the axioms representing the domain topology, axioms are introduced to capture topologically inconsistent relations. During a post processing step, individuals participating in the latter type of axioms are iteratively removed. Compared to the approach to inconsistency handling presented in this paper, [60] does not address the semantics of expressive constructors, while neither the implementation details of this postprocessing step nor evaluation results are given.

In [30], DLs and rules have been utilised for video annotation using crisp semantics. Additionally, there is no mentioning to what happens in the case of inconsistency. In [61], a DLs based approach to medical image annotation is presented under the assumption of crisp, consistent analysis extracted descriptions. In [62] a reasoning approach adhering to fuzzy logic principles was investigated for the purpose of integrating image descriptions extracted by means of visual analysis and textual analysis, regarding user entered descriptions, while in a more recent study presented in [33], a fuzzy DLs based reasoning framework has been proposed for the enhancement of initial descriptions acquired through statistical classifiers. As aforementioned, the presented approach extends on the last two investigations, building upon the acquired experiences.

Fuzzy DLs have been proposed in [58] for the purpose of semantic multimedia retrieval; the fuzzy annotations however are assumed to be available. Similar considerations have been investigated in older works such that of [63], where again the usefulness and significance of multimedia content descriptions that reflect the uncertainty present is pointed out. In the context of analysis, fuzzy DLs have been only recently explored in [55], where fuzzy DLs reasoning is used to infer semantic concepts based on part-of relations and to subsequently merge at image level. Possible inconsistencies in the analysis extracted description on which part-whole reasoning is employed is not addressed. Fuzzy logic semantics have also been investigated in [64] under a different application context, namely

---

[7] This is a direct result of treating the concepts to be inferred as aggregates of simpler ones and an interpretation as the quest of those aggregated and simple assertions that once introduced make use of the analysis extracted assertions.

for the purpose of supporting personalised information retrieval. In contrast to the approach presented in this paper, the emphasis is placed on weighted fuzzy concepts that are used to represent user preferences and contextualised preferences in order to allow the ranking of retrieved documents with respect to their relevance.

8   Conclusions and Future Work

The richness of visual information and the growth in the volume made available, render the potential for the exploitation of image content tremendous. Although the role of machine learning in the extraction of image semantics continues to grow, the reported endeavours show that the weakness to effectively incorporate semantics bears significant limitations in terms of the number of concepts that can be supported and the robustness of the attained performance. At the same time, the utilisation of explicit semantics as means to partially alleviate and enhance descriptions extracted through statistical learning presents an appealing potential, as suggested by recent studies addressing both research and industrial aspects [37,65].

Utilising fuzzy DLs semantics, the proposed reasoning framework captures the uncertainty of the extracted descriptions and accomplishes their integrated interpretation, while resolving inconsistencies rising from contradictory descriptions. In addition, by means of logical entailment, the final interpretation is further enriched; thereby, the need for training classifiers for semantically related concepts is partially alleviated, while missing descriptions due to segmentation and classification errors can be partially compensated. Experimentation has shown promising results, that although not conclusive yet, suggest that the proposed framework has the potential to serve as a useful contribution.

As indicated earlier in the paper, the investigation of a reasoning framework that combines fuzzy and probabilistic reasoning constitutes a challenging direction for future work. The motivation issues from the fact that the two types of uncertainty serve complementary purposes, hence suggesting a strong potential for achieving mutual benefit. However, more immediate directions for future investigations constitute on one hand on the extension of the presented reasoning framework so as to handle spatial knowledge, as well as the formalisation of the proposed reasoning tasks based on the drawn experiences with respect to the aspects that render the typical DL services inappropriate for direct exploitation in the problem of semantic image interpretation. Finally, towards more conclusive observations, we plan to extend our experimentation to larger, public data sets.

Acknowledgements

This work was partially supported by the European Commission under contracts FP6-001765 aceMedia, FP6-507482 KnowledgeWeb and FP7-215453 WeKnowIt.

References

1. Smeulders, A.W.M., Worring, M., Santini, S., Gupta, A., Jain, R.: Content-based image retrieval at the end of the early years. IEEE Trans. Pattern Anal. Mach. Intell. 22(12), 1349–1380 (2000)
2. Chang, S.F.: The holy grail of content-based. IEEE MultiMedia 9(2), 6–10 (2002)
3. Naphade, M., Huang, T.: Extracting semantics from audio-visual content: the final frontier in multimedia retrieval. IEEE Transactions on Neural Networks 13(4), 793–810 (2002)
4. Hanjalic, A., Lienhart, R., Ma, W., Smith, J.: The holy grail of multimedia information retrieval: So close or yet so far away. IEEE Proceedings, Special Issue on Multimedia Information Retrieval 96(4), 541–547 (2008)
5. Burges, C.: A Tutorial on Support Vector Machines for Pattern Recognition. Data Mining and Knowledge Discovery 2(2), 121–167 (1998)
6. Heckerman, D.: A tutorial on learning with bayesian networks. Learning in Graphical Models, 301–354 (1998)
7. Chapelle, O., Haffner, P., Vapnik, V.N.: Support vector machines for histogram-based image classification 10(5), 1055–1064 (1999)
8. Naphade, M., Huang, T.: A probabilistic framework for semantic video indexing, filtering, and retrieval. IEEE Transactions on Multimedia 3(1), 141–151 (2001)
9. Assfalg, J., Bertini, M., Colombo, C., Bimbo, A.D.: Semantic annotation of sports videos. IEEE MultiMedia 9(2), 52–60 (2002)
10. Christmas, W.J., Jaser, E., Messer, K., Kittler, J.: A multimedia system architecture for automatic annotation of sports videos. In: ICVS, pp. 513–522 (2003)
11. Town, C., Sinclair, D.: A self-referential perceptual inference framework for video interpretation. In: Crowley, J.L., Piater, J.H., Vincze, M., Paletta, L. (eds.) ICVS 2003. LNCS, vol. 2626, pp. 54–67. Springer, Heidelberg (2003)
12. Snoek, C., Worring, M., van Gemert, J., Geusebroek, J., Smeulders, A.: The challenge problem for automated detection of 101 semantic concepts in multimedia. In: Proc. 14th ACM International Conference on Multimedia, Santa Barbara, CA, USA, October 23-27, pp. 421–430 (2006)
13. Rao, A., Jain, R.: Knowledge representation and control in computer vision systems. IEEE Expert, 64–79 (1988)
14. Crevier, D., Lepage, R.: Knowledge-based image understanding systems: A survey. Computer Vision and Image Understanding 67, 161–185 (1997)
15. Snoek, C., Huurnink, B., Hollink, L., Rijke, M., Schreiber, G., Worring, M.: Adding semantics to detectors for video retrieval. IEEE Transactions on Multimedia 9(5), 975–986 (2007)
16. Horrocks, I., Patel-Schneider, P., van Harmelen, F.: From shiq and rdf to owl: the making of a web ontology language. J. Web Sem. 1(1), 7–26 (2003)
17. Horrocks, I., Patel-Schneider, P., Bechhofer, S., Tsarkov, D.: Owl rules: A proposal and prototype implementation. J. Web Semantics 3(1), 23–40 (2005)
18. Baader, F., Calvanese, D., McGuinness, D.L., Nardi, D., Patel-Schneider, P.F.: The description logic handbook: Theory, implementation, and applications. In: Description Logic Handbook. Cambridge University Press, Cambridge (2003)
19. Baader, F., Horrocks, I., Sattler, U.: Description logics as ontology languages for the semantic web. In: Mechanizing Mathematical Reasoning, pp. 228–248 (2005)
20. Hunter, J.: Adding Multimedia to the Semantic Web: Building an MPEG-7 Ontology. In: Proc. The First Semantic Web Working Symposium (SWWS), California, USA (July 2001)

21. Simou, N., Saathoff, C., Dasiopoulou, S., Spyrou, E., Voisine, N., Tzouvaras, V., Kompatsiaris, I., Avrithis, Y., Staab, S.: An ontology infrastructure for multimedia reasoning. In: Proc. International Workshop on Very Low Bitrate Video Coding (VLBV), Sardinia, Italy, September 15-16, pp. 51–60 (2005)
22. Arndt, R., Troncy, R., Staab, S., Hardman, L., Vacura, M.: COMM: Designing a well-founded multimedia ontology for the web. In: Aberer, K., Choi, K.-S., Noy, N., Allemang, D., Lee, K.-I., Nixon, L.J.B., Golbeck, J., Mika, P., Maynard, D., Mizoguchi, R., Schreiber, G., Cudré-Mauroux, P. (eds.) ASWC 2007 and ISWC 2007. LNCS, vol. 4825, pp. 30–43. Springer, Heidelberg (2007)
23. Dasiopoulou, S., Tzouvaras, V., Kompatsiaris, I., Strintzis, M.G.: Capturing mpeg-7 semantics. In: Proc. International Conference on Metadata and Semantics (MTSR), Corfu, Greece, October 11-12 (2007)
24. Troncy, R.: Integrating structure and semantics into audio-visual documents. In: Fensel, D., Sycara, K., Mylopoulos, J. (eds.) ISWC 2003. LNCS, vol. 2870, pp. 566–581. Springer, Heidelberg (2003)
25. Hunter, J., Drennan, J., Little, S.: Realizing the hydrogen economy through semantic web technologies. IEEE Intelligent Systems Journal - Special Issue on eScience 19, 40–47 (2004)
26. Petridis, K., Bloehdorn, S., Saathoff, C., Simou, N., Dasiopoulou, S., Tzouvaras, V., Handschuh, S., Avrithis, Y., Kompatsiaris, I., Staab, S.: Knowledge representation and semantic annotation of multimedia content. IEE Proceedings on Vision Image and Signal Processing, Special issue on Knowledge-Based Digital Media Processing 153, 255–262 (2006)
27. Little, S., Hunter, J.: Rules-by-example – A novel approach to semantic indexing and querying of images. In: McIlraith, S.A., Plexousakis, D., van Harmelen, F. (eds.) ISWC 2004. LNCS, vol. 3298, pp. 534–548. Springer, Heidelberg (2004)
28. Moller, R., Neumann, B., Wessel, M.: Towards computer vision with description logics: Some recent progress. In: Proc. Workshop on Integration of Speech and Image Understanding, Corfu, Greece, September 21, pp. 101–115 (1999)
29. Neumann, B., Moller, R.: On scene interpretation with description logics, FBI-B-257/04 (2004)
30. Bagdanov, A., Bertini, M., DelBimbo, A., Serra, G., Torniai, C.: Semantic annotation and retrieval of video events using multimedia ontologies. In: Proc. IEEE International Conference on Semantic Computing (ICSC), Irvine, CA, USA, pp. 713–720 (2007)
31. Espinosa, S., Kaya, A., Melzer, S., Möller, R., Wessel, M.: Multimedia interpretation as abduction. In: Proc. International Workshop on Description Logics (DL), Brixen-Bressanone, Italy, June 8-10, pp. 323–331 (2007)
32. Dasiopoulou, S., Mezaris, V., Kompatsiaris, I., Papastathis, V., Strintzis, M.: Knowledge-assisted semantic video object detection. IEEE Trans. Circuits Syst. Video Techn. 15(10), 1210–1224 (2005)
33. Dasiopoulou, S., Kompatsiaris, I., Strintzis, M.: Using fuzzy dLs to enhance semantic image analysis. In: Duke, D., Hardman, L., Hauptmann, A., Paulus, D., Staab, S. (eds.) SAMT 2008. LNCS, vol. 5392, pp. 31–46. Springer, Heidelberg (2008)
34. Maron, O., Ratan, A.: Multiple-instance learning for natural scene classification. In: Proc. 15th International Conference on Machine Learning (ICML), Madison, Wisconson, USA, July 24-27, pp. 341–349 (1998)
35. Vailaya, A., Figueiredo, M., Jain, A., Zhang, H.: Image classification for content-based indexing. IEEE Transactions on Image Processing 10(1), 117–130 (2001)
36. Barnard, K., Duygulu, P., Forsyth, D., de Freitas, N., Blei, D., Jordan, M.: Matching words and pictures. Journal of Machine Learning Research 3, 1107–1135 (2003)

37. Hauptmann, A., Yan, R., Lin, W.H., Christel, M., Wactlar, H.: Can high-level concepts fill the semantic gap in video retrieval? a case study with broadcast news. IEEE Transactions on Multimedia 9(5), 958–966 (2007)
38. Niemann, H., Sagerer, G., Schröder, S., Kummert, F.: Ernest: A semantic network system for pattern understanding. IEEE Trans. Pattern Anal. Mach. Intell. 12(9), 883–905 (1990)
39. Reiter, R., Mackworth, A.K.: A logical framework for depiction and image interpretation. Artif. Intell. 41(2), 125–155 (1989)
40. Russ, T., MacGregor, R., Salemi, B., Price, K., Nevatia, R.: Veil: Combining semantic knowledge with image understanding. In: ARPA Image Understanding Workshop, Palm Springs, CA, USA, February 12-17 (1996)
41. Rabiner, L., Juang, B.: An introduction to hidden markov models. IEEE ASSP Magazine, [see also IEEE Signal Processing Magazine] 3(1), 4–16 (1986)
42. Dubois, D., Prade, H.: Possibility theory, probability theory and multiple-valued logics: A clarification. Annals of Mathematics and Artificail Intelligence 32(1-4), 35–66 (2001)
43. Zadeh, L.: Fuzzy sets. Information and Control 8(32), 338–353 (1965)
44. Klir, G., Yuan, B.: Fuzzy sets and fuzzy logic: Theory and applications. Prentice-Hall, Englewood Cliffs (1995)
45. Yen, J.: Generalizing term subsumption languages to fuzzy logic. In: Proc. 12th International Joint Conference on Artificial Intelligence (IJCAI), Sydney, Australia, August 24-30, pp. 472–477 (1991)
46. Straccia, U.: A fuzzy description logic. In: Proc. International Conference on Artificial Intelligence and 10th Innovative Applications of Artificial Intelligence Conference (AAAI/IAAI), Madison, Wisconsin, July 26-30, pp. 594–599 (1998)
47. Straccia, U.: Reasoning within fuzzy description logics. J. Artif. Intell. Res. (JAIR) 14, 137–166 (2001)
48. Straccia, U.: Transforming fuzzy description logics into classical description logics. In: Alferes, J.J., Leite, J. (eds.) JELIA 2004. LNCS (LNAI), vol. 3229, pp. 385–399. Springer, Heidelberg (2004)
49. Stoilos, G., Stamou, G., Tzouvaras, V., Pan, J., Horrocks, I.: The fuzzy description logic f-SHIN. In: International Workshop on Uncertainty Reasoning For the Semantic Web (URSW), Galway, Ireland, November 7, pp. 67–76 (2005)
50. Stoilos, G., Stamou, G., Pan, J.: Handling imprecise knowledge with fuzzy description logic. In: Proc. International Workshop on Description Logics (DL), Lake District, UK, pp. 119–127 (2006)
51. Bell, D., Qi, G., Liu, W.: Approaches to inconsistency handling in description-logic based ontologies. In: Meersman, R., Tari, Z., Herrero, P. (eds.) OTM-WS 2007, Part II. LNCS, vol. 4806, pp. 1303–1311. Springer, Heidelberg (2007)
52. Lam, J., Sleeman, D., Pan, J., Vasconcelos, W.: A fine-grained approach to resolving unsatisfiable ontologies. In: Spaccapietra, S. (ed.) Journal on Data Semantics X. LNCS, vol. 4900, pp. 62–95. Springer, Heidelberg (2008)
53. Straccia, U.: A fuzzy description logic for the semantic web. In: Sanchez, E. (ed.) Fuzzy Logic and the Semantic Web. Capturing Intelligence, pp. 73–90. Elsevier, Amsterdam (2006)
54. Stoilos, G., Stamou, G., Pan, J., Tzouvaras, V., Horrocks, I.: Reasoning with very expressive fuzzy description logics. J. Artif. Intell. Res. (JAIR) 30, 273–320 (2007)
55. Simou, N., Athanasiadis, T., Tzouvaras, V., Kollias, S.: Multimedia reasoning with f-shin. In: 2nd International Workshop on Semantic Media Adaptation and Personalization (SMAP), London, UK, pp. 413–420 (2007)

56. Bobillo, F., Straccia, U.: fuzzydl: An expressive fuzzy description logic reasoner. In: Proc. International Conference on Fuzzy Systems (FUZZ), Hong Kong, June 1-6, pp. 923–930. IEEE Computer Society, Los Alamitos (2008)

57. Papadopoulos, G.T., Mylonas, P., Mezaris, V., Avrithis, Y., Kompatsiaris, I.: Knowledge-assisted image analysis based on context and spatial optimization (2006)

58. Umberto, S., Giulio, V.: Dlmedia: an ontology mediated multimedia information retrieval system. In: Proc. International Workshop on Description Logics (DL), Brixen-Bressanone, Italy, June 8-10, pp. 467–475

59. Neumann, B., Weiss, T.: Navigating through logic-based scene models for high-level scene interpretations. In: Crowley, J.L., Piater, J.H., Vincze, M., Paletta, L. (eds.) ICVS 2003. LNCS, vol. 2626, pp. 212–222. Springer, Heidelberg (2003)

60. Schober, J.P., Hermes, T., Herzog, O.: Content-based image retrieval by ontology-based object recognition. In: Proc. KI 2004 Workshop on Applications of Description Logics (ADL), Ulm Germany, September 24, pp. 1–10 (2004)

61. Hu, B., Dasmahapatra, S., Lewis, P., Shadbolt, N.: Ontology-based medical image annotation with description logics. In: Proc. 15th IEEE International Conference on Tools with Artificial Intelligence (ICTAI), Sacramento, California, USA, November 3-5, pp. 77–83 (2002)

62. Dasiopoulou, S., Heinecke, J., Saathoff, C., Strintzis, M.: Multimedia reasoning with natural language support. In: Proc. IEEE International Conference on Semantic Computing (ICSC), Irvine, CA, USA, September 17-19 (2007)

63. Meghini, C., Sebastiani, F., Straccia, U.: A model of multimedia information retrieval. J. ACM 48(5), 909–970 (2001)

64. Mylonas, P., Vallet, D., Castells, P., Fernandez, M., Avrithis, Y.: Personalized information retrieval based on context and ontological knowledge 23(1), 73–100 (March 2008)

65. Leger, A., Heinecke, J., Nixon, L., Shvaiko, P., Charlet, J., Hobson, P., Goasdoue, F.: Semantic web take-off in a european industry perspective. In: Garcia, R. (ed.) Semantic Web for Business: Cases and Applications, ch. 1, pp. 1–29. IGI Global (2008)

MISM: A Platform for Model-Independent Solutions to Model Management Problems

Paolo Atzeni, Luigi Bellomarini, Francesca Bugiotti, and Giorgio Gianforme

Dipartimento di informatica e automazione
Università Roma Tre
{atzeni}@dia.uniroma3.it,
{bellomarini,bugiotti}@yahoo.it,
{giorgio.gianforme}@gmail.com

Abstract. Model management is a metadata-based approach to database problems aimed at supporting the productivity of developers by providing schema manipulation operators.

Here we propose MISM (Model Independent Schema Management), a platform for model management offering a set of operators to manipulate schemas, in a manner that is both model-independent (in the sense that operators are generic and apply to schemas of different data models) and model-aware (in the sense that it is possible to say whether a schema is allowed for a data model). This is the first proposal for model management in this direction.

We consider the main operators in model management: merge, diff, and modelgen. These operators play a major role in solving various problems related to schema evolution (such as data integration, data exchange or forward engineering), and we show in detail a solution to a major representative of the class, the round-trip engineering problem.

Keywords: model management, model management operators, round-trip engineering, model-independent schema and data translation.

1 Introduction

The need for complex transformations of data arises in many different contexts, because of the presence of multiple representations for the same data or of multiple sources that need to coexist or to be integrated [11,18,20]. A major goal of technology in the database field is to enhance the productivity of software developers, by offering them high-level features that support repetitive tasks. This has been stressed since the introduction of the relational model, with the emphasis on set-oriented operations [12,13], but it was pursued, at least implicitly, in earlier developments of generalized techniques [22]. The *model management* proposal [7,8] is a recent, significant effort in this direction: its goal is the development of techniques that consider metadata and operations over them. More precisely, a model management system [11] should handle schemas and mappings between them by means of operators supporting operations to discover correspondences between

S. Spaccapietra, L. Delcambre (Eds.): Journal on Data Semantics XIV, LNCS 5880, pp. 133–161, 2009.
© Springer-Verlag Berlin Heidelberg 2009

schemas (MATCH), performing the most common set-oriented operations (such as union of schemas, MERGE, and difference of schemas, DIFF) and translating them from a data model to another (MODELGEN). These operations should be specified at a high level, on schemas and mappings, and should allow for the (support to the) generation of data-level transformations. Many application areas can benefit from the use of model management techniques, including data integration over heterogeneous databases, data exchange between independent databases, ETL (Extract, Transform, Load) in data warehousing, wrapper generation for the access to relational databases from object-oriented applications, dynamic Web site generation from databases.

Most of the work in model management has considered the need for *model independence*, that is, the fact that the techniques do not refer to individual data models,[1] but are more general. In detail, this requires that a single implementation of the operators should fit (i.e. be applicable) to any schema regardless of the specific data model it belongs to. This has usually been done by adopting some "universal data model," a model that is more general than the various models of interest in a heterogeneous framework. In the literature, such a data model is called *universal metamodel* [11] or *supermodel* [3,6]. If the operations of interest also include translations from a data model to another (the MODELGEN operator), it is important that the individual data models are represented, in such a way that it becomes possible to describe the fact that a schema belongs to a data model. We will call this property *model-awareness*. The various proposals for MODELGEN [3,6,25,26] do include the model independence feature, to a larger or lesser extent. For the other operators, the major efforts in the model management area (as summarized by Bernstein and Melnik [11]) do not handle the explicit representation of data models nor generic definitions of the operators.

The goal of this paper is to show a model independent and model aware approach to model management, thus providing concrete details to Bernstein's original proposal [8] and contributing to support its feasibility.

In the rest of this introductory section we first discuss two motivating examples, then we provide an overview of the approach and finally we state the contribution of the paper and the organization of the rest of it.

1.1 Motivating Examples

In order to have a context for specific examples and a complete solution, we will refer to the "round-trip engineering" problem [8], which can be defined as follows: given two schemas, where the second is somehow obtained from the first (for example, generated in a semiautomatic way, with standard rules partially overridden by human intervention), the problem has the goal of "repairing" the first if the second is modified. This problem is often considered in model management papers [8] as a representative of the "schema evolution" family. These problems arise in all application settings and therefore can be used to

---

[1] There is some disagreement on terminology in the literature: we use the term *data model* here for what is often called just *model* [3,6] and in other papers *metamodel* [11].

demonstrate the effectiveness of model management, in terms of both individual operators and compositions of them.

Let us consider an example derived from an academic scenario (see Figure 1): a university has various schools and one of them has a relational database with a portion containing all the information of interest about its departments, courses, and professors. Its schema is shown in the box labeled S_1 in Figure 1. It is composed of three tables, *Professor*, *Course*, and *Department*. Apart from the specific attributes, each relation has a key, denoted by the "ID" suffix and underlined in the figure. As each course is offered by a specific department and given by a professor, there are foreign keys from *Course* to the other two tables, denoted by arrows in the figure.

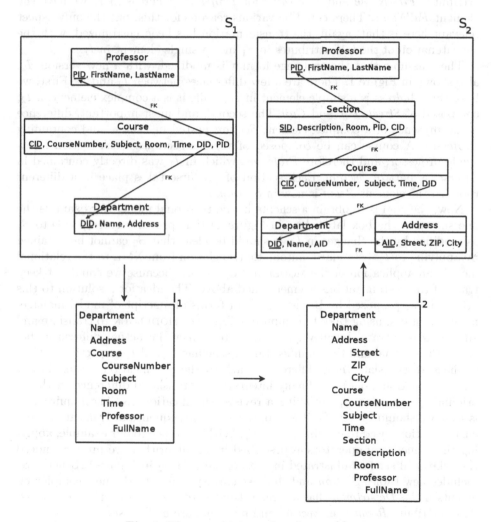

Fig. 1. The round-trip engineering problem

Assume now that this portion of the database is used (together with other goals) as the source to send data on courses to a central office in the university, which gathers data from all schools. This office requires data in an XML format, which is the one sketched in the box labeled I_1 in Figure 1. There is indeed a close correspondence between S_1 and I_1 (possibly because they were designed together). In fact, I_1 can be obtained by means of a nesting operation based on departments, each with the associated set of courses and with the instructor for each course. Clearly, this is one natural way to transform the relational data in S_1 into XML, but not the only one, as there would be other solutions that involve course or professor as the root. In this sense, we can say that this is not the result of an automated translation, but of a customization, that is, a choice among a few standard alternatives. Let us also observe that in S_1 we have attributes *FirstName* and *LastName* for *Professor*, whereas in I_1 we have the element *FullName*. There could be various reasons for this, but the only aspect relevant here is that, again, the transformation has been customized, with the concatenation of the two attributes in S_1 into a single element in I_1.

Then, assume that the exchange format is modified, with a new version, I_2, also shown in Figure 1. There are a few differences between I_2 and I_1. First, we have that *Address* is a simple element in I_1, while it is a complex element in I_2, composed of *Street*, *Zip*, and *City*. The second, and most important, difference is the presence of a complex element *Section* nested in *Course* and containing *Professor*. A course can be composed of various sections. Each section has a single professor, and therefore *Professor*, which in I_1 was directly contained in *Course*, is part of *Section*. Each section of a course takes place in a different room so the element *Room* is now in *Section*.

Now, the goal is to obtain a schema in the relational model (for example the one shown in the box labeled S_2 in Figure 1) that properly corresponds to S_1 as modified by the changes in I_2. It should be clear that S_2 cannot be obtained by applying to I_2 a standard, automatic translation from XML to the relational model (an application of the MODELGEN operator), because we could not keep track of the customizations we mentioned above. The idea for a solution to this problem was proposed by Bernstein [8], in terms of a script of model management operators, using DIFF to compute differences, MODELGEN to translate and MERGE to integrate. Intuitively, we have to detect the actual differences between the original and the modified target schemas I_1 and I_2 respectively. Then we have to translate these differences back to the specification model (in our case the relational one) and finally integrate the translated differences with the original specification S_1 obtaining a revised specification S_2. The requirement is that we should obtain I_2, if we apply to S_2 the sequence of translations and customizations used to obtain I_1 from S_1. With reference to our example, applying the sequence of operators as described in the algorithm, we produce indeed the relational schema illustrated in the box labeled S_2 in Figure 1. Schema S_2 includes new tables *Section* and *Address* corresponding to the new complex elements in I_2. *Department* has a foreign key to *Address* and *Section* to *Course*. Also, attribute *Room* is in *Section* and not anymore in *Course*.

In the existing literature, the proposals for the various operators are not general and accurate enough, as they refer to a rather limited set of models and do not have features that support the description of models, and so the plan proposed by Bernstein has not yet been implemented in a general way.

The goal of this paper is to show that this plan can indeed be made concrete, in a model-independent and model-aware way, which works for many different models but performs the translations knowing the specific features of the models of interest.

With the twofold goal of using a different model and of presenting a simpler example, let us consider another scenario. Let us assume we have a high level specification tool that translates ER schemas into relational tables by generating appropriate SQL DDL, allowing some customization. Again, if changes are made to the SQL implementation, then we want them to be propagated back to the ER specification. This is illustrated in Figure 2, where S_1 represents a specification in the ER model and I_1 represents its relational implementation. The customization in the translation produces two columns *FName* and *LName* in I_1 for the single attribute *Name* in S_1. Then, if I_1 is modified to a new version I_2, the latter is not coherent with S_1. The main difference between I_2 and I_1 is in the key for the

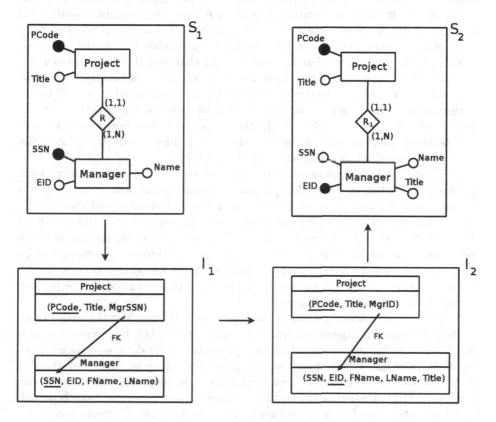

Fig. 2. A simple scenario for the round-trip engineering problem

Manager table and, as a consequence, in the foreign key structure that refers to it. Also, *Manager* has a new attribute, *Title*. The goal is to find a specification S_2 from which I_2 could be generated, in the same semiautomatic way as I_1 was obtained from S_1. Indeed, what we want to obtain is an ER schema S_2, which differs from the original one in the attributes of the entity *Manager*: the identifier is *EID* instead of *SSN* and there is the new attribute *Title*.

In the remainder of the paper we will follow this second example, which will allows us to explain completely our approach, without taking too much space.

1.2   Overview of the Approach

The solution we propose in this paper includes a definition and implementation of the major model management operators (DIFF, MERGE, and MODELGEN). It is based on our experience in the MIDST platform [3,4,5], where a model-independent approach for schema and data translation was introduced (with a generic implementation of the MODELGEN operator). MIDST adopts a met-alevel approach in which the artifacts of interest are handled in a repository that represents data models, schemas, and databases in an integrated way, both model-independent and model-aware. This is a fundamental starting point, as stated before, in order to be able to define a model management system. This repository is implemented as a multilevel dictionary. Data models are defined in terms of the constructs they involve. A schema of a specific data model is allowed to use only the constructs that are available for that model. In this framework, the *supermodel* is the model that includes the whole range of constructs, so that every schema in every model is also a schema in the supermodel. Then, all translations are performed within the supermodel, in order to scale with respect to the size of the space of models [5]. In this paper, we show how the dictionary and the supermodel provide grounds for the model-independent definition of the other operators of interest, namely MERGE and DIFF.

In MIDST, translations are obtained as the composition of basic steps each of which is written as a Datalog program. The language was chosen for two reasons: first, it matches in an effective way the structure of our data model and dictionary (which is implemented in relational form); second, its high level of abstraction and the declarative form allow for a clear separation between the translations and the engine that executes them. Moreover, Datalog can be straightly translated into SQL and the original choice was aimed at covering the widest spectrum of application scenarios. However, other syntax or specification formalisms could be adopted as well.

Here we propose a general model management platform, MISM (Model Independent Schema Management), which is based on MIDST but extends it in a significant way. We start from MIDST's representation for data models, schemas, and databases and define model management operators by means of Datalog programs with respect to such representation. Specifically, we leverage on the features of MIDST's dictionary for the uniform representation of models as well as the infrastructure for the definition and the application of schema manipulation operators. MISM offers all the major operators, including MERGE, DIFF, and a basic version of MATCH, all implemented in a model-generic way. The structure of the dictionary also allows for the automatic generation of Datalog programs implementing the new operators, with respect to the given supermodel, in such a way that, if the supermodel were extended, the operators would be automatically extended as well.

1.3 Contribution

To the best of our knowledge, this is the first proposal for a model-independent platform for model management. Specifically, this paper offers three main contributions:

- The model-independent definition and implementation of important model management operators. In fact, we define them by means of programs with predicates acting on the constructs of the supermodel.
- The automatic generation of the programs implementing the operators only using the supermodel as input. These programs are valid for any schema defined in terms of model-generic constructs.
- A complete solution to the round-trip engineering problem as a representative of the problems that can be solved with this approach. It is based on a script defined in terms of a convenient combination of our operators and allows a walk through of our implementation.

1.4 Organization of the Paper

In Section 2 we describe how models, schemas and translations are dealt with in MIDST. In particular we describe schema representation within MIDST metalevel. We illustrate how model-independent transformations can be performed in the framework.

Then in Section 3 we illustrate in detail model management operators in MISM, the extension of MIDST we propose here, and present their definitions. Discussion on their model-independence and model-awareness is provided. As a consequence, in Section 4 we show possible Datalog implementations for these operators satisfying the specifications of the previous section.

Section 5 presents a solution of the round-trip engineering problem in terms of our operators and shows how MISM can be used to solve this problem. A concrete scenario of solution, addressing the problem introduced in Figure 2 is then provided.

Finally, Section 6 discusses related work and Section 7 concludes the paper.

2   Models, Schemas, and Translations in MIDST

This section presents the needed background, with a discussion of the relevant features of our previous project, MIDST [3,5], whose goal was to provide a generic version of the MODELGEN operator, which can be defined as follows:

given a source schema S expressed in a source model, and a target model TM, MODELGEN generates a schema S' expressed in TM that is "equivalent" (according to a suitable definition) to S. MIDST obtains model-independence and model-awareness by means of the adoption of a rich dictionary, which stores models, schemas and data in a uniform and coordinated way. In this paper, we leverage on MIDST from two points of view: first, we show definitions and implementations of additional operators, MERGE and DIFF, and it is again the organization of the dictionary that supports model-independence and model-awareness; second, MODELGEN is used in the scripts we propose, together with the new operators. Hence both MIDST dictionary and its implementation of the MODELGEN operator are part of the new model management platform we propose in this paper.

MIDST adopts a model-generic representation of schemas based on a combination of constructs. Its founding observation is the similarity of features which arises across different data models. This means that all the existing models can be represented with a rather small set of general purpose constructs [21] called *metaconstructs* (or simply *constructs* when no ambiguity arises). Let us briefly illustrate this idea. Consider the concept of entity in the ER model family and that of class in the OO world: they both have a name, a collection of properties and can be in some kind of relationship between one another. To a greater extent, it is easy to generalize this observation to any other construct of the known models and determine a rather small set of general constructs. Therefore models are defined as sets of constructs from a given universe, in which every construct has a specific name (such as "entity" or "object"): for instance a simple version of the ER model may be composed of Abstracts (the entities), Aggregations of Abstracts (the relationships) and Lexicals referring to Abstracts (attributes of entities); instead the relational model could have Aggregations (the tables), Lexicals referring to Aggregations (the columns), and foreign keys specified over finite sets of Lexicals. Thus schemas are collections of actual constructs (schema elements) related to one another. Figure 3 lists the metaconstructs used in the current version of MIDST [5] and the corresponding specific constructs we have in various popular (families of) data models.

As we said in the introduction, the set of all the possible constructs in MIDST forms the *supermodel*, a major concept in our framework. It represents the most general model, such that any other model is a specialization of it (since a subset of its constructs). Hence a schema S of a model M is necessarily a schema of the supermodel as well.

MIDST manages the information of interest in a rich dictionary. Its details have been described elsewhere [2] and are beyond the scope of this paper. Let us summarize its main features. It has two layers, both implemented in the relational model: a *basic* level and a *metalevel*.

The basic layer of the dictionary has a model-specific part (some tables of which are shown in Figure 4 with reference to our running example), where schemas are represented with explicit reference to the various models, and, more important, a model-generic one, where there is a table for each construct in the supermodel:

Metaconstruct	Relational	Object-Relational	ER	XSD
Abstract	-	typed table	entity	root element
Lexical	column	column	attribute	simple element
BinaryAggregationOf-Abstracts	-	-	binary relationship	-
AbstractAttribute	-	reference	-	-
Generalization	-	generalization	generalization	-
Aggregation	table	table	-	-
ForeignKey	foreign key	foreign key	-	foreign key
StructOfAttributes	-	structured column	-	complex element

Fig. 3. Simplified representation of MIDST metamodel

ER_ENTITY

OID	Entity-Name	Schema
e1	Project	s1
e2	Manager	s1
...	...	...

ER_ATTRIBUTEOFENTITY

OID	Entity	Att-Name	Type	isKey	Schema
a1	e1	PCode	int	true	s1
a2	e1	Title	string	false	s1
a3	e2	SSN	int	true	s1
a4	e2	EID	int	false	s1
a5	e2	Name	string	false	s1
...	...	...	...	...	...

ER_BINARYRELATIONSHIP

OID	Rel-Name	Entity1	IsOptional1	IsFunctional1	Entity2	...	Schema
b1	R	e1	false	true	e2	...	s1
...	...	...	...	...	...	...	...

REL_TABLE

OID	Table-Name	Schema
t1	Project	i1
t2	Manager	i1
...	...	...

REL_COLUMN

OID	Table	Col-Name	Type	isKey	Schema
c1	t1	PCode	int	true	i1
c2	t1	Title	string	false	i1
...	...	...	...	...	i1
c7	t2	LName	string	false	i1
...	...	...	...	...	...

Fig. 4. A portion of a model-specific representation of schemas S_1 and I_1 of Figure 2

so there is a table for SM_ABSTRACT (the SM_ prefix emphasizes the fact that we are in the supermodel portion of the dictionary), a table for SM_AGGREGATION and so on (with an example in Figure 5). These tables have a column for each property of interest for the construct (for example, a Lexical can be part of the identifier of the corresponding Abstract, or not, and this is described by means of a boolean property). References are used to link constructs to one another, and

SM_ABSTRACT		
OID	Abs-Name	Schema
e1	Project	s1
e2	Manager	s1
...	...	...

SM_LEXICAL					
OID	Abstract	Aggr	Lex-Name	Type	isId	Schema
a1	e1	-	PCode	int	true	s1
a2	e1	-	Title	string	false	s1
a3	e2	-	SSN	int	true	s1
...	...	...	...	...	...	...
c1	-	t1	PCode	int	true	i1
...	...	...	...	...	...	...
c7	-	t2	LName	string	false	i1
...	...	-	...	...	...	...

SM_AGGREGATION		
OID	Aggr-Name	Schema
t1	Project	i1
t2	Manager	i1
...	...	...

SM_BINARYAGGREGATIONOFABSTRACTS							
OID	Agg-Name	Abstract1	IsOptional1	IsFunctional1	Abstract2	...	Schema
b1	R	e1	false	true	e2	...	s1
...	...	...	...	...	...	...	...

Fig. 5. A portion of a model-generic representation of the schemas S_1 and I_1 of Figure 2

so the tables in the dictionary have fields with foreign keys connecting them to each other. For example, the SM_LEXICAL table has an attribute that contains references to SM_ABSTRACT, to represent the fact that a Lexical (for example an attribute of entity in the ER model) has to belong to a parent construct, which could be an Abstract (an entity). In both parts, constructs are organized in such a way they guarantee the *acyclicity constraint*, meaning that no cycles of references are allowed between them. This is convenient in situations where a complete navigation through the schemas is necessary and a topological order is helpful.

The two parts of the dictionary play complementary roles in the translation process, which is MIDST's main goal: the model specific part is used to interact with source and target schemas and databases, whereas the supermodel part is used to perform translations, by referring only to constructs, regardless of how they are used in the individual models. This allows for model-independence.

In fact, every translation in MIDST is composed of three phases: first, the source schema, expressed in a specific source model, is copied into the supermodel; second, the actual translation is carried out in the supermodel environment; finally, the result schema, which refers to the supermodel, but is compatible with the target model, is copied into the target model itself. The translation engine exploits a library of elementary translations, each of which is written as a Datalog program, and combines them, on the basis of the specific source and target model of interest.

MIDST dictionary includes a higher layer, a *metalevel*, which gives a characterization of the construct properties and relationships among them [2,5]. It involves few tables, each with few rows, which form the core of the dictionary. A significant portion is shown in Figure 6. Its main table, named MSM_CONSTRUCT (here, the MSM_ prefix denotes that we are in the "metasupermodel" world, as we

MSM-CONSTRUCT		
OID	Construct-Name	IsLex
mc1	Abstract	false
mc2	Lexical	true
mc3	BinaryAggregationOfAbstracts	false
mc4	AbstractAttribute	false
...	...	...

MSM-PROPERTY			
OID	Prop-Name	Constr	Type
mp1	Abstract-Name	mc1	string
mp2	Att-Name	mc2	string
mp3	IsId	mc2	bool
mp4	IsFunctional1	mc3	bool
mp5	IsFunctional2	mc3	bool
...	...	...	...

MSM-REFERENCE			
OID	Ref-Name	Constr	ConstrTo
mr1	Abstract	mc2	mc1
mr2	Abstract1	mc3	mc1
mr3	Abstract2	mc3	mc1
...	...	...	...

Fig. 6. The supermodel part of the metalevel portion of the dictionary of MIDST

are describing the supermodel) stores the name and a unique identifier (OID) for each construct, so this table actually memorizes every allowed construct; indeed, the rows of this table correspond essentially to those in Figure 3. Each construct is also characterized by a set of properties describing the details of interest. There is a table, MSM_PROPERTY, reporting name, type and owner construct for each property. The properties, for example, allow to define whether an entity attribute is identifier or not and to specify the cardinality of relationships. Constructs refer to one another with references, recorded in the table MSM_REFERENCE.

As we have illustrated, the metalevel lays the basis for the definition of constructs which can be then used in defining models and so on the structure of the lower layer of the dictionary: in fact, the model-generic layer (Figure 5) has one table for each row in MSM_CONSTRUCT (and so we have, as we said, tables named SM_ABSTRACT, SM_AGGREGATION, SM_LEXICAL, and so on), with columns corresponding to the properties and references of the construct, as described in MSM_PROPERTY and MSM_REFERENCE, respectively.

The aim of the following sections is to define operators on the basis of constructs in such a way that model-independent solutions to model management problems can then be described. In fact, solutions will be formulated as scripts involving the application of such operators. We will see that the structure of the dictionary, especially with its metalevel, plays a major role in the automatic generation of Datalog programs for the implementation of the operators.

3 Operators

Model management, as we said in the introduction, refers to a wide range of problems, which share the need for high level solutions. Therefore many operators have been proposed, depending on the family of problems of interest. Here we concentrate on schema evolution, where proposals [8,10] require MATCH, DIFF and MERGE and, if an explicit representation of models is needed, also MOD-ELGEN. In such proposals, the MATCH operator is used to discover mappings between the elements of the involved schemas. In fact, mappings play a major role, as they provide the operators with essential information about the relationships between the involved schemas. For example, an operator that computes the difference between two schemas needs to know the correspondences between constructs in order to subtract them correctly. Likewise, an operator that combines schemas must know those correspondences in order to avoid the generation of duplicates. Here, exploiting our construct-based representation of data models, we can propose definitions of the main operators (DIFF, MERGE, and MODELGEN) that compare constructs on the basis of their names and structures. In fact, we assume that if two constructs have different names or different structures, they should be considered as different. In this way, as we clarify in the next subsection, our approach considers MATCH as complementary.

We already have an implementation for MODELGEN in our MIDST proposal (and hence in MISM as well), and so we have to concentrate on DIFF and MERGE. In the rest of this section we will present specifications for these operators that refer to MIDST dictionary, preceded by the discussion of a preliminary notion, equivalence of schema elements. Then, in Section 4 we will show how to generate Datalog implementations for them.

3.1 Equivalence of Schema Elements

The basic idea behind the DIFF and MERGE operators is the set-theoretical one. In fact, we can consider each schema as composed of a set of *schema elements* (the actual constructs it involves), and then consider DIFF as a set-theoretic difference (the elements that are in the first schema and not in the second) and MERGE as a union (the elements that are at least in one of the two schemas). In general, we might be interested in comparing schemas that represent the concepts of interest by means of different elements. In such a case, a preliminary step would require the identification or specification of the correspondences between them. This is usually done by means of an application of the MATCH operator, which, in general, can produce correspondences of various types (i.e. one-to-one, one-to-many, or even many-to-many) and may require a human intervention in order to disambiguate or to better specify. Besides, in MIDST context, let us recall that each schema element is represented with respect to a specific model-generic construct (i.e. an element refers to an Abstract, another one refers to an Aggregation and so on): in this sense we say that an element is an *instance of* a construct. Consequently, we distinguish between *construct-preserving* correspondences and *non construct-preserving* ones. The first type maps elements, instances of a certain construct, only to elements that are instances of the same model-generic construct; viceversa, correspondences not satisfying this property belong to the second type. For example in the XML schemas of Figure 1 the correspondence between the simple element *Address* and the complex one (again called *Address*), composed of *Street*, *Zip*, and *City*, is not construct-preserving. In fact the address is represented by a simple element in the first schema (i.e. a Lexical in MIDST), while in the second one it requires a complex element (i.e. a StructOfAttributes in MIDST) with its components (i.e. some Lexicals in MIDST). Clearly, non construct-preserving correspondences denote different ways to organize the data of interest and therefore the involved constructs of the two schemas have to be considered as different. On the other hand, constructs that have different names but the same structure while handling the same data, have to be considered as equivalent. These are one-to-one correspondences, which can be discovered manually or by means of a matching system (among the many existing ones [27]).

The arguments above lead to a notion of renaming of a schema: given a correspondence *c*, the *renaming* of a schema *S* with respect to *c* is a schema where the names of the elements in *S* are modified according to *c*. Then, we have a basic idea of equivalence conveyed by the following recursive statement:

two schema elements are equivalent with respect to a renaming if: (i) they are instances of the same model-generic construct; (ii) their names are equal, after the renaming; (iii) their features (names and properties) are equal; and (iv) they refer to equivalent elements.

For the sake of simplicity, we can assume that the renaming is always applied to one of the schemas, in order to guarantee that corresponding constructs with the same type also have the same name. In some sense this would correspond to a *unique name assumption*. Then, equivalence would be simpler, as name equality would be required:

two schema elements are equivalent if their types, names and features are equal and they refer to equivalent elements.

It is important to observe that the definition is recursive, as equivalence of pairs of elements requires the equivalence of the elements they refer to. This is well defined, because the structure of references in our supermodel is acyclic, and therefore recursion is bounded. Let us consider few cases from our running example, namely schemas I_1 and I_2 in Figure 2. We have a column *Title* for a table *Project* in both schemas, and the two are equivalent, as they have the same name, the same properties (they are both non-key), and refer to equivalent elements (the tables named *Project*). Instead, the column *Title* of *Project* in I_1 is not equivalent to *Title* of *Manager* in I_2, because *Project* and *Manager* are not equivalent. Also, the two columns named *SSN* are not equivalent, because the one in I_1 is key and that in I_2 is not.

3.2   Definitions of the Operators

We are now ready to give our definitions and show some examples. According to what we said in the previous section, we assume that suitable renamings have been applied in such a way that a unique name assumption holds. We start with a preliminary notion, to be revised shortly.

Given two schemas S and S', the difference DIFF(S, S') *is a schema S'' that contains all the schema elements of S that do not appear in S'.*

This first intuitive idea must be refined, otherwise some inconsistencies could arise. In fact, it may be the case that a schema element appears in the result of a difference while an element it refers to does not. This causes incoherent schemas with "orphan" elements. With respect to the schemas in our running example, this happens for the column *MgrID* in the difference DIFF(I_2, I_1), which belongs to the result, while the table *Project* does not. Instead we want to have *coherent* schemas, where references are not dangling.

In order to solve this difficulty, we modify our notion of a schema, by introducing *stub elements* (similar to the *support objects* of [8]). Specifically, we extend the notion of *schema element*, by allowing two kinds: *proper elements* (or simply *elements*), those we have seen so far, and *stub elements*, which are essentially fictitious elements, introduced to guarantee that required references exist. We say that a schema is *proper* if all its elements are proper.

According to this technique, the result of DIFF(I_2, I_1) contains the stub version of *Project* in order to avoid the missing reference of *MgrID*.

The definition of the difference should therefore be modified in order to take into account stub elements both in the source schemas and in the result one.

Given S and S', DIFF(S, S') *is a schema S'' that contains: (i) all the schema elements of S that do not appear in S'; (ii) stub versions for elements of S that appear also in S' (and so should not be in the difference) but are referred to by other elements in* DIFF(S, S').

The notion is recursive, but well defined because of the acyclicity of our references.

In the literature [8], the DIFF operator is often used in model management scripts to detect which schema elements have been added to or removed from a schema. Our definition addresses this target. Given an "old" schema S and a "new" one S', the "added" elements (also called the *positive* difference) can be obtained as DIFF(S', S) whereas the "removed" ones (the *negative* difference) are given by DIFF(S, S').

With respect to the running example in Figure 2, the negative difference, DIFF(I_1, I_2), contains the columns *MgrSSN* of *Project* and *SSN* (key) and *EID* (non-key) of *Manager*. Column *MgrSSN* belongs to the difference since it belongs to I_1 and there is no attribute with the same name in I_2. Instead, *EID* and *SSN* belong to DIFF(I_1, I_2) because the attributes with the same respective names in I_2 have properties that differ from those in I_1: *EID* is key in I_1 and not key in I_2, whereas the converse holds for *SSN*. The negative difference does not contain the two tables as proper elements, because they appear in both schemas, but it needs them as stub elements because the various columns have to refer to them. The negative difference also includes the foreign key in I_1 since it does not appear in I_2 (the foreign key in I_2 involves different columns).

Similarly, the positive difference includes the columns *MgrID* of *Project* and *SSN* (non-key), *EID* (key) and *Title* of *Manager*, both tables as stub elements, and the foreign key in I_2.

An important observation is that the definition we have given here is model-independent, because it refers to constructs as they are defined in our super-model. At the same time, it is *model-aware*, because it is always possible to tell whether a schema belongs to a model, on the basis of the types of the involved schema elements. As a consequence, it is possible to introduce a notion of closure: we say that a model management operator O is *closed with respect to a model M* if, whenever O is applied to schemas in M, then the result is a schema in M as well. Given the various definitions, it follows that the difference is a closed operator, because it produces only constructs that appear in its input arguments.

Let us now turn our attention to the second operator of interest, MERGE. We start again with a preliminary definition.

Given S and S', their merge MERGE(S, S') is a schema S'' that contains the schema elements that appear in at least one of S or S', modulo equivalence.[2]

It is clear that merge is essentially a set-theoretic union between two schemas, with the avoidance of duplicates managed by means of the notion of equivalence of schema elements.

Since our schemas might involve stub elements, as we saw above, let us consider their impact on this operator. Clearly, the operator cannot introduce new stub elements, as it only copies elements. However, stubs can appear in the input schemas, and the delicate case is when equivalent elements appear in schemas, proper in one and stub in the other.[3]

Given S and S', their merge MERGE(S, S') is a schema S'' that contains the schema elements that appear in at least one of S or S', modulo equivalence. An element in S'' is proper if it appears as proper in at least one of S and S' and stub otherwise.

As an example, consider the following schemas, each composed of a single table. S: *Project(PCode, Title)* and S': *Project(PCode, MgrSSN)*. Their merge will be another schema S'' containing the table *Project(PCode, Title, MgrSSN)*. Notice that the table *Project* and the column *PCode* appear both in S and in S'

---

[2] Technically, both here and in the difference, we should note that schema elements have their identity. Therefore, in all cases we have new elements in the results; so, here, we copy in the result schema the elements of the two input schemas, and "modulo equivalence" means that we collapse the pairs of elements of the two schemas that are equivalent (only pairs, with one element from each schema, as there are no equivalent elements within a single schema).

[3] Equivalence of elements neglects the difference between stub and proper elements, as it is not relevant in this context.

and, since they are recognized as equivalent, there are no duplicates in S''. The column *Title* appears only in S while *MgrSSN* only in S'; therefore one copy of each is present in the result schema S''. We will see a complete example of MERGE in Section 5, while discussing the details of our running example.

For this operator, arguments for model independence and model closure can be made in the same way as we did for DIFF: specifically, only schema elements deriving from schemas S and S' will appear in the result and, consequently, if they belong to a given model, then S'' will belong to that model as well.

For the sake of homogeneity in notation, let us define also the operator that performs translations between models:

Given a schema S of a source model M and a target model M', the translation MODELGEN(S, M') *is a schema S' of M' that corresponds to S.*

We have discussed at length MODELGEN elsewhere [3,5]. Here we just mention that this notation refers to a generic version of it that works for all source and target models (the source model is not needed in the notation as it can be inferred from the source schema), thus avoiding different operators for different pairs of models. Indeed, our MIDST implementation [4,5] of MODELGEN includes a feature that can select the appropriate translation for any given pair of source and target models.

4   Model-Independent Operators in MISM

In this section we show how the definitions of the operators can be made concrete, in a model-independent way, in our tool, leveraging on the structure of its dictionary. The implementation has been carried out in Datalog, and here we concentrate on its main principles, namely the high-level declarative specification, and the possibility of automatic generation of the rules, on the basis of the metalevel description of models.

The Datalog specification of each operator is composed of two parts:[4]

1. equivalence test;
2. procedure application.

The first part tests the equivalence to provide the second part with necessary preliminary information on the elements of the input schemas.

We first illustrate how the equivalence test can be expressed in Datalog, and then proceed with the discussion for the specific aspects of DIFF and MERGE. At the end of the section, we discuss how all these Datalog programs can be automatically generated out of the dictionary.

4.1   Equivalence Test

The first phase involves the implementation of a test for equivalence of constructs, according to the definition we gave in Section 3. Given the definition,

---

[4] For the sake of readability we describe them in a procedural way, even if the specification is clearly declarative.

all we need is a rule for each model-generic construct that compares the schema elements that are instances of such a construct. It refers to two schemas, denoted by the "schema variables" SOURCE_1 and SOURCE_2, respectively. It generates an intensional predicate (a view, in database terms) that indicates the pairs of OIDs of equivalent constructs. As an example, let us see the Datalog rule that compares Aggregations.

```
EQUIV_Aggregation [DEST] (OID1: oid1, OID2: oid2)
  <- SM_Aggregation [SOURCE_1] (OID: oid1, Name: name),
     SM_Aggregation [SOURCE_2] (OID: oid2, Name: name);
```

Aggregation has no references (and also no properties) and so the comparison is based only on name equality (verified with the variable *name*). If the names of the two Aggregations are equal, then they are equivalent, and so their OIDs are included in the view for equivalent Aggregations. In the running example, tables *Project* and *Manager* of the two schemas are detected as equivalent since they have the same names, respectively.

The situation becomes slightly more complex when constructs involve references. This is the case for Lexicals of Aggregation (in the running example, the various columns of *Project* and *Manager*).

```
EQUIV_Lexical [DEST] (OID1: oid1, OID2: oid2)
  <- SM_Lexical [SOURCE_1] (OID: oid1, Name: name, isIdentifier: isId,
         isNullable: isNull, type: t, aggregationOID: oid3),
     SM_Lexical [SOURCE_2] (OID: oid2, Name: name, isIdentifier: isId,
         isNullable: isNull, type: t, aggregationOID: oid4),
     EQUIV_Aggregation (OID1: oid3, OID2: oid4);
```

The first and the second body predicates compare names and homologous properties of a pair of Lexicals, one belonging to I_1 (SOURCE_1) and the other to I_2 (SOURCE_2). Comparisons are made by means of repeated variables (such as *name, isId, isNull, t*). Moreover, as Lexicals involve references to Aggregations (as no column exists without a table, in the example), we need to compare the elements they refer to. The last predicate in the body performs this task by verifying that the Aggregations (tables) referred to by the Lexicals (columns) of I_1 and I_2 are equivalent (i.e. the corresponding pair of OIDs is in the equivalence view for Aggregation).

If the constructs under examination belonged to a deeper level, there would be a predicate to test the equivalence of ancestors for each step of the hierarchical chain. Each predicate would query the appropriate equivalence view to complete the test. Termination is guaranteed by the acyclicity of the supermodel.

Let us observe that the Datalog program generated in this way is model-aware since it takes into account the type of constructs when performing comparisons. In fact, as it is clear in the examples, Datalog rules are defined with specific respect to the type of the constructs to be compared: a Lexical is compared only with another Lexical and so for an Abstract or other constructs.

The program is model generic as well, since the set of rules contains a rule for each construct in the supermodel. Then a given pair of schemas will really make use of a subset of the rules, the ones referring to the constructs they actually involve according to their model.

4.2   The DIFF Operator

The DIFF operator is implemented by a Datalog program with the following steps:

1. equivalence test (comparison between the input schemas);
2. selective copy.

The first step is the equivalence test we have described in Section 4.1.

As for the second step, there is a Datalog rule for each construct of the super-model, hence taking into account each kind of schema element: the rule verifies whether the OID of an element of the first schema belongs to a tuple in the equivalence view. If this happens, this means that there is an equivalent construct in the second schema, implying that the difference must not contain it, otherwise the copy takes place. For example, the rule for Aggregations results as follows:

```
SM_Aggregation [DEST] (OID: #AggregationOID_0(oid), Name: name)
    <- SM_Aggregation [SOURCE_1] (OID: oid, Name: name),
       !EQUIV_Aggregation (OID1: oid);
```

In the rule, the # symbol denotes a Skolem functor, which is used to generate new identifiers (in the same way as we did in MIDST [5]). Indeed, the functor is interpreted as an injective function, in such a way that the rule produces a new construct for each different source construct on which it is applicable. The various functions also have disjoint ranges. The rule copies into the result schema all the Aggregations of SOURCE_1 that are not equivalent to any Aggregation of SOURCE_2. The condition of non-equivalence is tested by the negated predicate (negation is denoted by "!") over the equivalence view; in fact, if the OID of an Aggregation of the first source schema is present in the view, then it has a corresponding Aggregation in the second source schema, and so it must not belong to the difference.

With reference to the running example, let us compute $\mathrm{DIFF}(I_1, I_2)$. The rule above represents the computation of the difference with respect to tables. Since in Figure 2 both *Project* and *Manager* in I_1 have an equivalent table in I_2, then the difference does not contain any Aggregation.

Consider now the rule for Lexicals (columns):

```
SM_Lexical [DEST] (OID: #LexicalOID_0(oid), Name: name,
        isIdentifier: isId, isNullable: isNull, type: t,
        aggregationOID: #AggregationOID_0(oid1))
    <- SM_Lexical [SOURCE_1] (OID: oid, Name: name,
        isIdentifier: isId, isNullable: isNull, type: t,
        aggregationOID: oid1),
       !EQUIV_Lexical (OID1: oid);
```

It copies into the result schema all the Lexicals of SOURCE_1 that are not equivalent to any Lexical of SOURCE_2. In the example of Figure 2, the Lexical *MgrSSN* has been removed from *Project*. Also, *SSN* of *Manager* is key in I_1 but not in I_2 and the converse for *EID*. Consequently all of the mentioned Lexicals will belong to the difference DIFF (I_1, I_2).

For the sake of simplicity, we have omitted from the above rules the features that handle stub elements. However the actual implementation of the difference requires them in order to address the consistency issues we have discussed in the previous section. The strategy we adopt is the following: when a non-first level element (that is, one with references) is copied, the procedure copies its referred elements too if they are not copied for another reason. Then, unless they are proper parts of the result, the procedure marks the referred elements as stub. The following rule exemplifies this with respect to Aggregations.

```
SM_Aggregation [DEST] (OID: #AggregationOID_0(oid), Name: name,
        isStub: true)
  <- SM_Aggregation [SOURCE_1] (OID: oid, Name: name),
     EQUIV_Aggregation (OID1: oid, isStub: false);
```

If a Lexical (referring to an Aggregation) belongs to the difference, then the referred Aggregation must be copied into the difference as stub (if it has not been copied directly). The rule above copies from the first schema every Aggregation that would not belong to the difference since it has an equivalent (non stub) element in the second schema (which is verified by the predicate over the view, which also contains information on whether the equivalence involves stub elements) and marks it as stub. As for the input, we must subtract schemas with stub elements properly. Thus the selective copy in step 2 must be adapted: it should copy (into the result schema) a non-stub element in the first schema only if the second schema does not contain a non-stub equivalent element. This last condition is tested by a predicate over an equivalence view like the one in the above Datalog rule.

The techniques described refer to the rules for the specification of the difference of schemas. Indeed, as our dictionary includes also a data level (as illustrated in a previous paper of ours [2]), which lists all data items that instantiate a given construct, it is interesting to see how the operator could be specified in such a way that the result is a schema, as we saw above, together with the associated data. While working at MODELGEN, we tackled the same issue, and we developed a technique that generated data level Datalog programs out of schema level ones [3]. In such a context, correctness was a delicate issue, as each translation has its own specific features, and the tool administrator has the responsibility of verifying the correctness. Here we are interested in a general program, that implements difference, and therefore we cannot rely upon the approval of a human. However, things are indeed easier, as the difference needs to include all instances of the constructs that appear in the result schema: for example, if the result of DIFF includes table *Manager*, then we need all its instances in the result database, but this is just a copy, as *Manager* is a table in the source schema as well. So, data level rules for DIFF could be produced as rules that copy all instances of constructs, with the condition that the construct appears in the result schema, which is easy to express, as it is indeed the condition in the body of the schema rule. Therefore, while we omit the details for the sake of space, we can safely claim that we can generate correct rules that operate on data from those that operate on schemas.

4.3 The MERGE Operator

The approach we follow for MERGE is based on the same ideas as the one for DIFF. We code it in terms of Datalog rules defined over the constructs of MIDST supermodel. Rules copy elements of one type to elements of the same type and we guarantee the needed model closure.

The MERGE operator, as defined in Section 3.2, is represented by a Datalog program with the following tasks:

1. equivalence test (comparison between the input schemas);
2. selective copy from the first argument;
3. selective copy from the second argument.

The first step involves the computation of an equivalence view containing the correspondences between the elements of the input schemas.

Assume we are computing $S'' = \text{MERGE}(S, S')$. In step 2 the procedure copies into the destination schema S'' all the elements in S, except those that are stub in S and non-stub in S'. In step 3 the procedure copies all the elements of S' that are not present in S and those that are non-stub in S' and stub in S.

The combination of these two steps implies that in S'' there will not be duplicates of any element. If an element is present both in S and S', in S as a stub and in S' as a non-stub, it will be present in S'' as a non-stub. A stub element will appear in the result as stub as well, if an element is present only in S or S' as a stub or both in S and S' as stub.

In such an implementation of the MERGE, a thorough handling of references is important and we achieve this by means of Skolem functions, which are injective as we said in the previous section. In fact, it may happen for an element of the result schema to have a stub parent in the first source schema and a non-stub parent coming from the second source schema: let E be an element of S which is copied into the result schema. E has a stub parent P in S and there is another element P' which is the equivalent non-stub element of P in S'. P will not be copied from S, but there will be its equivalent P' coming from S'. As a consequence, the reference of E to P must use an OID that is derived from the OID of P' in S' and not from the OID of P in S. As we have seen for the difference, this logic can be implemented in Datalog on the basis of a predicate over the equivalence views.

By following arguments similar to those for DIFF, we can claim that, from the schema level Datalog programs for MERGE, we can generate programs that implement the operator on data, thus performing the merge of the actual databases (in the internal representation in our dictionary). The reason is that the operator is again a sort of selective copy.

4.4   Automatic Generation of Datalog Programs for the Operators

The implementations of both the phases of the operators are based on comparisons and copies of schema elements considered in terms of constructs of the supermodel. We have seen in Section 2 that MIDST handles the descriptions of these constructs in a dictionary, defining their names, features and references to one another. An automatic generation of the Datalog programs we have presented is possible and indeed represents a key point of the approach we propose here. Concretely, we propose a new module of MISM, *OpGen*, that automatically generates the rules according to the supermodel constructs. OpGen reads the information in the dictionary about constructs, their references, and their properties, and uses it to produce appropriate Datalog rules in the right order, according to the structure of constructs. As we said in the respective sections, for each operator we can generate data level rules that perform the selective copy of the instances of the involved constructs.

Automatically generated operators are not only model-independent but also supermodel-independent. In fact, in case of extensions to and modifications of the supermodel, all we need is to use OpGen to generate an updated version of the operators.

It is worth noting that our model-generic operators are scalable, since their internal complexity does not depend on the size of the input schemas nor on the number of modifications. In fact, they are generated by OpGen once and work for every possible set of input schemas defined in terms of constructs of MIDST supermodel. Moreover, although more efficient implementations of them could be designed, their application is entirely devoted to the database system which addresses, as a consequence, all the optimization issues.

5   A Model-Independent Solution to the Round-Trip Engineering Problem

In the previous sections we described the most common model management operators. We have shown that since they are defined over the constructs of MIDST supermodel, they are model-independent; moreover we have shown that it is possible to exploit their model awareness in order to satisfy the model closure property. This implies that solutions to model management problems, given in terms of these operators, are model-independent.

Here we show how our approach can be used to provide a model-independent solution to the round-trip engineering problem, illustrated in the introduction as one of the most representative ones in the model management area.

5.1   The General Procedure

Consider Figure 7: S_1 is the *specification schema* and I_1 the *implementation* schema obtained from S_1 with the application of the transformation (a translation and, possibly, some customizations) map_1. Let I_2 be a modified version of I_1. The goal is to determine a specification S_2 from which I_2 could be derived.

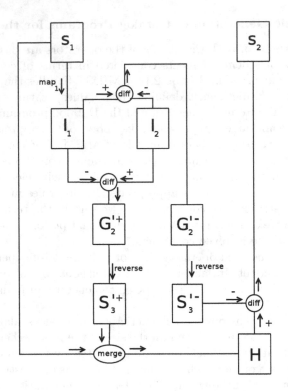

Fig. 7. A procedure for the round-trip engineering problem

Operationally, we assume that I_1 has been generated from the specification schema by the MODELGEN operator, possibly followed by a customization step; viceversa, we make no specific assumption on how I_2 has been obtained: it could be some transformation (specified by means of a Datalog program or in some other way), or a manual modification or evolution of I_1, or it could even come from an external input.

Then the procedure is as follows.

1. $G_2'^- = \text{DIFF}(I_1, I_2)$
   Here we use the DIFF operator to detect which elements of the implementation schema I_1 do not appear in the revised version I_2: these are the elements belonging to I_1 but not to I_2 (i.e. the removed elements).

2. $G_2'^+ = \text{DIFF}(I_2, I_1)$
   This difference (with parameters swapped with respect to the previous one) allows to compute which elements have been added in the revision which led from I_1 to I_2. In fact, these elements are all the ones present in I_2 but not in I_1.

3. $S_3'^-$ is obtained by applying to $G_2'^-$ the reverse of the mapping map_1. The details then depend on the way map_1 is defined. In the common case where it is an automatic translation from the specification model to the implementation one (an application of MODELGEN), possibly followed by a customization, we

have that reverse can be done with MODELGEN as well, with a translation from the implementation model to the specification one. This ignores the possible customizations, under the assumption that changes in I_1 (yielding I_2) do not involve customized elements. In fact, if this is the case, $G_2'^-$ will not include the customized elements, since they are removed by the difference step. It should be noted that in general the existence of the inverse of a given translation is not guaranteed. We will discuss this issue later in this section.

4. Similarly for the other difference: $S_3'^+$ is obtained by applying to $G_2'^+$ the reverse of the mapping map_1.

5. $H = \text{MERGE}(S_1, S_3'^+)$
   H is the union of the original specification S_1 with the reversed difference $S_3'^+$ containing the added elements. Therefore, H contains all the original elements plus the added ones.

6. $S_2 = \text{DIFF}(H, S_3'^-)$
   The last operation of the procedure subtracts $S_3'^-$ from the temporary result H, because the elements in $S_3'^-$ are those that correspond to the elements removed in the implementation.

It is clear that this procedure does not require information about the models of the source schemas, since the operators act at MISM metalevel, dealing with constructs directly, however the model awareness of MISM guarantees the model closure. In fact, in the same way as we do for translations in our previous tool MIDST (see Section 2), we apply our operators in the supermodel framework, and the procedure is preceded and followed by copy steps, the first from the specific source model to the supermodel and the second from the supermodel to the specific model, which essentially rename constructs. An example should get the meaning across: suppose the specification data model is ER, while the implementation belongs to the relational model. Before applying the DIFF between I_1 and I_2, we rename all the elements in terms of constructs of the supermodel. After this step, there is no need to take into account the model-specific constructs anymore and the procedure can continue with respect to model-generic constructs only. Then, since the operators are defined in such a way that the difference between two schemas of a model belongs to that model, then we are guaranteed that the two differences in the procedure belong to the relational model as well. Finally, we apply MERGE and DIFF on ER schemas. These operators work independently of the model. However, we are sure that the results will also belong to the ER model because, as we have illustrated, the operators do not add any new element.

Moreover, it is important to observe that, if S_1, I_1 and I_2 are proper (and coherent,[5] as we always assume) schemas, then the result S_2 of the script is a proper schema as well. Consider the last two steps of the procedure ((i) $H = \text{MERGE}(S_1, S_3'^+)$),-(ii) $S_2 = \text{DIFF}(H, S_3'^-)$): S_1 is assumed to be proper (the script starts from a specification without stubs). $S_3'^+$ contains added elements which

---

[5] As we said in Section 3.2, a schema is coherent if all its constructs have no dangling references to other constructs.

may refer to stub parents. However, as I_1 and I_2 are coherent, we have that non-stub equivalents for these stub parents are already present in S_1. Therefore H is proper. $S_3'^-$ contains the removed constructs. Then, as I_2 is coherent, in $S_3'^-$ we cannot come across the removal of parent elements when their descendants are preserved. Therefore S_2 is proper.

In the above procedure, we have referred to applications of MODELGEN from the specification model to the implementation one and viceversa, as if they were one the inverse of the other. This need not be always the case, because models have different expressive power. However, from the practical point of view, we have reasonable solutions, as follows. A preliminary observation is that our translations can be seen as schema mappings where the correspondences are represented by Skolem functions. In general, schema mappings are not always invertible according to the strict definition, but in the literature there are proposals for relaxed constraints guaranteeing the existence of a kind of inverse mapping. According to Fagin et al. [15] a Local As View (LAV) schema mapping, having a set of Tuple Generating Dependencies (TGDs) where their left-hand sides are singleton, always admits a *quasi-inverse* corresponding mapping. Let us consider a mapping m and a source schema S; applying m to S we obtain another schema T. A quasi-inverse mapping does not permit to reobtain S (with its original data) from T, however, it allows to obtain a schema S^* such that applying m to it we have T again (with all its data). In our approach the only translation rules dealing with the actual data are the ones involving Lexicals. All these rules are LAV TGDs and therefore the whole translation is a LAV schema mapping and so each translation admits at least a quasi-inverse one that is part of the MISM repository. In general, a translation can lead to loss of information (i.e. when we translate a model into a less expressive one); in such cases it is not possible to define an inverse translation, but only a quasi-inverse one. It is worth noting that this loss of information has already been accepted by the user of the system when performing the first translation (from the specification to the implementation). Moreover, this is the only loss of information of the whole process. In fact after the first translation, it is possible to apply the quasi-inverse translation and the direct one repeatedly always obtaining the same schemas (with the same data). The inverse (quasi-inverse) translation does not cause loss of information even if it turns a model into a more expressive one. In fact, the input schema of the inverse translation has been obtained from a schema of a less expressive model; therefore it contains only structures that can be represented in such a model.

5.2   Application of the Round-Trip Solving Procedure

Now we present the details of the application of the round-trip solving procedure described in Subsection 5.1 to the case already shown in Figure 2. The specification domain is the ER model, while the implementations are relational schemas. It is a common scenario in which high level specifications are conceptually designed with an ER schema. The implementation, which in this situation belongs to the relational model, is then derived from the ER through the application of a translation rule.

The various steps are shown in Figure 8. Schema S_1 is composed of two entities, *Project* and *Manager*, and has a relationship R between them. *PCode* and *Title* are *Project* attributes (*PCode* is key), while *SSN*, *Name* and *EID* are *Manager* attributes (*SSN* is key).

Map_1 is implemented in two parts: a first part of the transformation is represented by ER-to-relational translation rule. A second part of it consists of the customization step which splits *Name* into *FName* and *LName*.

The transformation from the old to the new implementation modifies the table *Project* by changing the name of its column *MgrSSN* (to *MgrID*); it also modifies the *Manager* by adding the column *Title* and changing its key (from *SSN* to *EID*). The foreign key that in I_1 connects the column *SSN* with the table *Manager*, does not exist anymore, it is replaced by a new foreign key from the column *MgrID* of *Project* to the table *Manager*.

The first step of the solving procedure is the double application of the DIFF rules to I_1 and I_2 which yields $G_2'^-$ (negative difference) and $G_2'^+$ (positive difference), as we have already seen with examples for the operator in Section 3.2.

Then each semi-difference is reversed with the application of the MODELGEN operator, with the ER model as a target. In the case under examination, the reverse translation is simple, while in general it might be much more complex. Notice that in the application of the reverse rule, the stubness property of elements is preserved, then for example the entity *Project* in $S_3'^+$ is stub as well as in $G_2'^+$. Notice that the foreign key of $G_2'^-$ is reversed into the relationship R (that is the same as in S_1,[6] while the foreign key of $G_2'^+$ is reversed into the relationship R_1 (that is different from the one in S_1).

Now we have three different versions of the specification: the original one, S_1, together with $S_3'^-$, including all the elements that have to be removed, and $S_3'^+$, containing all the added elements.

The set-oriented merge of schemas S_1 and $S_3'^+$ leads to an updated specification, H, containing all the initial elements plus the added ones. Then in H we have *Project* with *PCode* (coming from S_1) and *Title* (from S_1) (the table *Project* is not stub anymore since it comes from S_1); moreover, there is the table *Manager* (non-stub for the same reason as *Project*) with the attributes *Name* (coming from S_1), *SSN* (from S_3^+), *SSN* (key) (from S_1), *EID* (from S_1), *EID* (key) (from S_3^+) and *Title* (from S_3^+). H also contains two relationships, R (coming from S_1) and R_1 (from $S_3'^+$).

Finally, we need to subtract from H all the non-stub elements in $S_3'^-$. Therefore, *SSN* (key) and *EID* are not present in the obtained result S_2. The relationship R of H is also present in $S_3'^-$, so the only relationship between *Project* and *Manager* in S_2 will be R_1.

---

[6] We can get back the "original" name because each construct has a name property; hence also the foreign key has a name property (not shown in figure) in our construct-based representation; in detail, we instantiated the name of the foreign key during the translation from S_1 to I_1 and we did the same during this step.

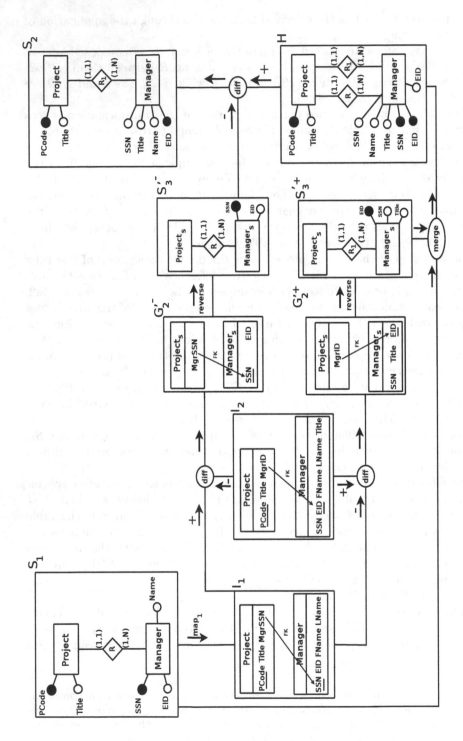

Fig. 8. An example of application of the round-trip solving procedure

6   Related Work

This paper illustrates a general approach to model management and relies on our previous work on model-generic schema and data translation [1,3,4,5] describing our conception and implementation of the MODELGEN operator. There are many proposals addressing model management problems which have been put forward since the original formulation of the problem.

In [7] Bernstein et al. recognize the possibility of a generic metadata approach to model management: their theoretical formalizations [8] and later studies converged into Rondo, a programming platform for model management [23]. However their approach is not supported by a description of models and so they pursue model independence without a concrete characterization of models and they cannot associate schemas with models. Conversely, MIDST (and now MISM) uses a dictionary of models and schemas to actually represent models and allows transparent transformations on them.

A parallel but orthogonal approach to model management problems, is that of Clio [16,17,19,24,28] whose aim is the development of a user aiding environment that allows the specification of a mapping between two instances and, consequently, generates the rules to implement the high level specified correspondences. Clio mainly offers a solution to data exchange problems by generating directly executable, though approximate, mappings between schemas. Similarly to Rondo, it lacks a model-independent representation of schemas and a representation of models.

A recent approach to schema evolution is PRISM [14]. Citing the authors, PRISM provides an intuitive, operational interface, used by the database administrator to evaluate the effect of possible evolution steps with respect to redundancy, information preservation, and impact on queries. In detail, the administrator can use a Schema Modification Operators (SMO) [9] language in order to specify schema changes and check whether such a modification could cause information loss, introduce redundancy, or grant invertibility. Moreover, the system allows for an automatic migration of the data, grants compatibility with old queries (i.e. against an old schema), and maintains the schema history. We propose something wider in which this approach can fit well: with reference to our running example, for instance, we could use similar techniques in order to constrain the evolutionary step between implementation schemas, thus granting the aforementioned desirable properties.

Our approach, together with Bernstein's, is more general and proposes a global platform for model management where the generation of executable mappings, like Clio's or PRISM's, is a complementary feature.

7   Conclusions

In this paper, we have discussed a paradigm and a concrete platform allowing model-independent solutions to a wide range of model management problems. We have provided effective definitions and implementations of model management operators which can be directly executed by the MISM platform. The operators defined in this way have been used to assemble a solution to major model management problems.

A major target of the model management research is the development of an advanced software system managing all the involved problems (model management system). Such a system aims at providing applications with an abstraction layer towards data programmability issues, that is, the whole spectrum of application problems concerning data manipulation. The approach presented in this paper lies in this direction. MIDST represents a framework for model management problems; MISM is an enhanced version, where operators and solving procedures are specifically designed to maximize the abstraction level together with an effective and sound representation of schemas and models. In parallel, we are working on the development of runtime strategies and algorithms in order to make our solutions in step with large operational databases as well as compliant with the most expressive data models.

Acknowledgement

We would like to thank the anonymous reviewers for their very helpful comments.

References

1. Atzeni, P., Cappellari, P., Bernstein, P.A.: Modelgen: Model independent schema translation. In: ICDE Conference, pp. 1111–1112. IEEE Computer Society, Los Alamitos (2005)
2. Atzeni, P., Cappellari, P., Bernstein, P.A.: A multilevel dictionary for model management. In: Delcambre, L.M.L., Kop, C., Mayr, H.C., Mylopoulos, J., Pastor, Ó. (eds.) ER 2005. LNCS, vol. 3716, pp. 160–175. Springer, Heidelberg (2005)
3. Atzeni, P., Cappellari, P., Bernstein, P.A.: Model-independent schema and data translation. In: Ioannidis, Y., Scholl, M.H., Schmidt, J.W., Matthes, F., Hatzopoulos, M., Böhm, K., Kemper, A., Grust, T., Böhm, C. (eds.) EDBT 2006. LNCS, vol. 3896, pp. 368–385. Springer, Heidelberg (2006)
4. Atzeni, P., Cappellari, P., Gianforme, G.: MIDST: model independent schema and data translation. In: SIGMOD Conference, pp. 1134–1136. ACM, New York (2007)
5. Atzeni, P., Cappellari, P., Torlone, R., Bernstein, P.A., Gianforme, G.: Model-independent schema translation. VLDB J. 17(6), 1347–1370 (2008)
6. Atzeni, P., Torlone, R.: Management of multiple models in an extensible database design tool. In: Apers, P.M.G., Bouzeghoub, M., Gardarin, G. (eds.) EDBT 1996. LNCS, vol. 1057, pp. 79–95. Springer, Heidelberg (1996)
7. Bernstein, P., Haas, L., Jarke, M., Rahm, E., Wiederhold, G.: Panel: Is generic metadata management feasible? In: VLDB Conference, pp. 660–662 (2000)
8. Bernstein, P.A.: Applying model management to classical meta data problems. In: CIDR Conference, pp. 209–220 (2003)
9. Bernstein, P.A., Green, T.J., Melnik, S., Nash, A.: Implementing mapping composition. VLDB J. 17(2), 333–353 (2008)
10. Bernstein, P.A., Halevy, A.Y., Pottinger, R.: A vision of management of complex models. SIGMOD Record 29(4), 55–63 (2000)

11. Bernstein, P.A., Melnik, S.: Model management 2.0: manipulating richer mappings. In: SIGMOD Conference, pp. 1–12. ACM, New York (2007)
12. Codd, E.: A relational model for large shared data banks. CACM 13(6), 377–387 (1970)
13. Codd, E.: Relational database: A practical foundation for productivity. CACM 25(2), 109–117 (1982)
14. Curino, C., Moon, H.J., Zaniolo, C.: Graceful database schema evolution: the PRISM workbench. PVLDB 1(1), 761–772 (2008)
15. Fagin, R., Kolaitis, P., Popa, L., Tan, W.: Quasi-inverses of schema mappings. ACM Trans. Database Syst. 33(2), 1–52 (2008)
16. Fagin, R., Kolaitis, P.G., Miller, R.J., Popa, L.: Data exchange: semantics and query answering. Theor. Comput. Sci. 336(1), 89–124 (2005)
17. Fagin, R., Kolaitis, P.G., Popa, L.: Data exchange: getting to the core. ACM Trans. Database Syst. 30(1), 174–210 (2005)
18. Haas, L.M.: Beauty and the beast: The theory and practice of information integration. In: Schwentick, T., Suciu, D. (eds.) ICDT 2007. LNCS, vol. 4353, pp. 28–43. Springer, Heidelberg (2006)
19. Haas, L.M., Hernández, M.A., Ho, H., Popa, L., Roth, M.: Clio grows up: from research prototype to industrial tool. In: SIGMOD Conference, pp. 805–810. ACM, New York (2005)
20. Halevy, A.Y., Ashish, N., Bitton, D., Carey, M.J., Draper, D., Pollock, J., Rosenthal, A., Sikka, V.: Enterprise information integration: successes, challenges and controversies. In: SIGMOD Conference, pp. 778–787. ACM, New York (2005)
21. Hull, R., King, R.: Semantic database modelling: Survey, applications and research issues. ACM Computing Surveys 19(3), 201–260 (1987)
22. McGee, W.C.: Generalization: Key to successful electronic data processing. J. ACM 6(1), 1–23 (1959)
23. Melnik, S.: Generic Model Management: Concepts and Algorithms. Springer, Heidelberg (2004)
24. Miller, R.J., Haas, L.M., Hernández, M.A.: Schema mapping as query discovery. In: VLDB Conference, pp. 77–88 (2000)
25. Mork, P., Bernstein, P.A., Melnik, S.: Teaching a schema translator to produce O/R views. In: Parent, C., Schewe, K.-D., Storey, V.C., Thalheim, B. (eds.) ER 2007. LNCS, vol. 4801, pp. 102–119. Springer, Heidelberg (2007)
26. Papotti, P., Torlone, R.: Heterogeneous data translation through XML conversion. J. Web Eng. 4(3), 189–204 (2005)
27. Rahm, E., Bernstein, P.A.: A survey of approaches to automatic schema matching. VLDB J. 10(4), 334–350 (2001)
28. Velegrakis, Y., Miller, R.J., Popa, L.: Mapping adaptation under evolving schemas. In: VLDB Conference, pp. 584–595 (2003)

Author Index